POLYMER CHEMISTRY
Synthesis and Characterization Techniques

Prashant D. Ashtaputrey
Santosh D. Ashtaputrey
Assistant Professor (Chemistry)
Government Institute of Science,
R.T. Road, Civil Lines,
NAGPUR (M.S.)

PRINTS PUBLICATIONS PVT. LTD.
NEW DELHI

Published by :

Prints Publications Pvt. Ltd.
Prints House, 11, Darya Ganj
New Delhi - 110002
Phone: +91-11-45355555
E-mail : contact@printspublications.com
Website: www.printspublications.com

First Edition : 2018

ISBN 978-81-927694-2-4

Price : ₹ 495/-

Published and Printed by Mr. Pranav Gupta on behalf of Prints Publications Pvt Ltd at Mohit Enterprises, Delhi.

CONTENTS

PREFACE

While conducting the practical of polymer science of post graduate students of chemistry, we found it very difficult to have a laboratory practical manual reference book which covers the experimental procedure for the synthesis and characterization of polymers. Therefore we have decided to write a book on experimental polymer chemistry to provide assistance to the students and teachers regarding the synthesis and characterization of polymers. We have tried to include the maximum content in the book so as to fulfill the need of comprehensive laboratory course in polymer science for undergraduate and postgraduate students of various courses like B.Tech, M. Tech, BE and M.Sc. in all Indian universities. The book may also cater the need of faculties working in the educational institutes.

The book is divided into three main sections. The first section deals with the basic laboratory safety measures and introduction to polymers. The classification, techniques of polymerization and characterization of polymers by different methods have been discussed in this section. There are various characterization techniques which are not possible to perform in the normal laboratories due to the lack of instrumental facilities like NMR, IR, XRD etc. In such cases we have tried to give the basic ideas of characterization or the information that we can get from these instrumental techniques. We have discussed the topics with assuming that the readers are well aware of the theoretical principles behind all of these instrumental techniques. The second section included the actual experimental procedures involved in the synthesis of various polymers. In the third section we have discussed the experimental characterization of polymers by various chemical and analytical techniques. We have explained all the experiments in a very lucid and simple way so that every student may understand it easily and perform well in the laboratory.

We are greatly thankful Dr. A.R. Raut, Associate professor (Chemistry) Institute of Science, Nagpur for his continuous encouragement for not only writing but throughout the preparation of this book. We wish to express our gratitude to Dr. R. G. Atram, Director, Institute of Science and Prof. V. T Kamble, Head, Department of Chemistry for their support and blessings. We also take opportunity to thank our colleagues Dr. S.S. Deo, Dr. G.M. Phadnaik, Shri. R.S. Shinde and Dr. R. Tayde for the useful discussions with them which were valuable in bringing out this book.

We are also thankful to Mr. Pranav Gupta, Director, Prints Publications Pvt. Ltd. for publishing this book within a short span of time in a well design format.

Since no work can be perfect and the improvement is a continuous process, therefore we shall feel obliged to have comments, healthy criticisms and valuable suggestions from the readers for the improvement of this book.

Nagpur

Authors
S. D. Ashtaputrey
P. D. Ashtaputrey

SECTION– I

1. Basic Laboratory Work Guidelines
2. Introduction to polymers
3. Characterization of polymers

1

Basic Laboratory Work Guidelines

Science laboratory is a place where the various scientific concepts and ideas can be verified by performing the experiments. Acquiring the scientific knowledge and scientific attitude are the two main objectives of working in science laboratory and can be achieved only by doing regular practical work in the laboratory. The achievements of the modern science are mainly due to the applications of the experimental methods. It is therefore important that the practical work should form a prominent feature in any science course. As a result of proper practical work the students cultivates many good habits like working skill, co-operation, resourcefulness, initiative, self dependence and confidence etc which are useful in their day to day life. The individual working in the laboratory is always preferred over the working in groups.

1.1. LABORATORY SAFETY

Safety is of the utmost concern in every chemical laboratory. Most of the chemicals used in laboratory are hazards. Therefore the students should be familiar with the safety issues related to working with chemicals in a chemistry laboratory, the following precautions must be taken while working in laboratory.

1. Students should work in the laboratory under the proper guidance of teacher only.

2. The experiments must be carried out in a fume hood.
3. Safety glasses with side shields must be worn at all times while working in the laboratory.
4. Students should be properly dressed and laboratory coats or apron must be worn at all times when in the laboratory.
5. The proper hand gloves must be worn whenever necessary while working in the laboratory.
6. Check the glass wares for any minor cracks before actually using them.
7. Experiments in progress must not be left unattended.
8. Noxious or odorous chemicals are to be handled only in a fume hood.
9. Never handle or pour flammable liquids near an open flame or other ignition source.
10. Pay attention to any hazards and warnings given in each experiment.
11. It is generally good practice to wash your hands frequently when working in any chemical laboratory.
12. At the end of the practical, make sure that the hood, work area and sink are clean and tidy.
13. Wash your hands before you leave.

1.2. LABORATORY ROUGH NOTEBOOK

It is important to keep the record of the procedure and observations while working in the laboratory in the notebook. This is even more important as a means of communication between students and the teacher. After a gap, interruptions, the record helps in how to proceed further with the work. The notes should help in detecting causes of unusual observations, as well as causes of accidents or fires. To be most effective, each student should have a clipboard posted near the work area. This clipboard will include all the information needed for the experiment as given below,

1. The title of the experiment.
2. Introduction including the brief summary of the experiment that is to be performed.

3. Notes or any special instructions from the teacher.
4. A table of physical constants of the materials that are to be used, of the products, co-products, and by-products.
5. Notes on the proposed disposal of the products and associated materials.
6. A step-by-step outline of the procedure of the work.
7. Calculations involved in the experiments.

Students should verify the results after the completion of experiment by the teacher with appropriate date and time and certify it by writing the experiment in the practical journal on next day.

1.3. LABORATORY SAFE PRACTICES

Accident may occur in the laboratory so it is essential to keep the first aid kit in the laboratory at easily accessible place. It must be in good condition and regularly inspected by the teacher in charge. The prevention is better than cure and therefore every precaution should be taken to avoid any type of accident in the laboratory. Some of the commonly occurring laboratory accidents and their remedies are given below.

(*i*) *Fire*

If the vapors from a flask ignite, a small fire can be extinguished by turning off the burner and gently placing wire gauze over the top of the vessel containing the burning solvent. Under any circumstances do not touch and move the flask. If solvent spread on bench tops ignites, if possible, move bottles and flasks of containing solvent away from the area and then use the fire extinguisher. If the person's clothes catch fire, then he or she should wrap tightly in a fire proof blanket.

(*ii*) *Explosions*

Explosions are very fast exothermic reactions and are usually given by substances which can undergo internal redox reactions e.g. polynitro compounds such as T.N.T. (trinitrotoluene), picric acid (trinitrophenol) or nitroglycerin, or by substances which can decompose to give much more stable compounds such as peroxides, azides, fulminates and diazo compounds. You will be warned if any particular explosion hazard exists with the compounds that you are using. One of the commonest causes of unexpected explosions is as a result of distilling solutions to dryness leaving behind traces of explosive residues such as peroxides. This is one reason why a distillation vessel is never heated to dryness.

(iii) Poisoning, Skin Contact, Chemicals in the Eyes

Most of the chemicals are poisonous e.g. most alcohols, amines and nitriles. Students should not try tasting, eating and drinking of chemicals in the laboratory. Do not pipette acids, liquids by mouth. You should always wash your hands before leaving the laboratory.

Certain harmful substances can be absorbed relatively easily through the skin e.g. dimethyl sulphate, nitrobenzene, aniline, phenol, and phenylhydrazine. Aromatic amines are serious carcinogens. Always minimise contact with any chemical and wash with water after accidental contact. You may wear latex surgical gloves, if you wish. These will be available in the laboratory, or for improved protection, wear a pair of heavy duty gloves.

If harmful, irritating or flammable gases are used or produced during a reaction, then that part of the experiment should be conducted in the fume-hood. If you feel irritating and burning into your eyes wash them thoroughly (for about 15 minutes) with plenty of water and contact the doctor immeadiately. Contact lenses must be removed immediately in order to wash the eyes.

Following are the some preliminary treatments essential in case of eye injuries

Acid in eye - The eye should be washed by continuous opening and closing under water. Then it should be washed with lime water or 1% solution of sodium bi carbonate.

Alkali in eye - The eye should be thoroughly washed with water and then with 1% boric acid solution.

HAZARD-Nitric acid, glacial acetic acid, potassium hydroxide, Hydrogen bromide and silver nitrate are corrosive. Chlorobenzene and chloroform are considered hazardous.

Organic solvents like benzene, acetone, alcohols, toluene etc are flammable and their vapors can be harmful. Precautions should be taken to avoid inhalation and skin or eye contact with these chemicals.

(*iv*) *Protection of Skin and Clothing*

Acids and alkalis are extremely corrosive and will damage skin and clothing. Certain organic compounds are excellent dyes. This is the reason that you are required to wear a protective laboratory coat and safety glasses. Sandals or shorts must not be worn in the laboratory. Beware of burns from hot glassware, hot iron rings or hot plates.

(*v*) *Mixing Chemicals*

Extreme caution should be used when mixing certain chemicals. If in doubt, ask your teacher. Be especially careful when dealing with waste as certain types of waste are incompatible (e.g. concentrated H_2SO_4 and organic waste).

Carefully observe the warnings about mixing water with certain reagents e.g. water should never be poured into concentrated H_2SO_4, or brought into contact with sodium. Chlorinated hydrocarbon solvent residues should be placed in the bottle provided. These should never be mixed with alkali. Never put concentrated H_2SO_4 in the organic waste containers.

(*vi*) *Handling Reagents*

Always replace bottle tops as soon as you have finished dispensing reagents since many compounds react with moisture in the air, with oxygen or with carbon dioxide. Others are quite volatile. Do not mix the tops. If you do remove the reagent from the common work area temporarily, replace it as soon as possible. Take great care to avoid contaminating reagents by putting the wrong pipette into the wrong reagent bottle. A reagent without proper label is useless and should not be used.

(*vii*) *Tidiness*

In order to work well with accuracy and efficiency, one of the first essentials is to keep your working place and the common areas such as reagent racks and fume hoods neat and clean. Wipe up any spillages immediately.

1.4. LABORATORY WASTE MANAGEMENT

This is very important while working in the laboratory. This is for your safety and to protect our environment. Appropriate waste disposal is important therefore make sure that, the waste is going into the appropriate container. Here are some guidelines for waste disposal,

- Never pour waste chemicals or reaction product down the sink other than very dilute aqueous, non-toxic solutions.
- Any broken glassware or other sharp objects should be collected carefully and should be place in a separate dust bin specifically marked for this purpose.
- Make sure that the waste is put in a proper container.
- Solid chemical waste should be placed in the container provided in the fume hood.
- Organic solutions or waste solvents should be placed in the organic waste container in the fume hood (different from the container for solid chemical waste). If possible the organic solvents can be carefully distilled from the waste solution for reuse.
- Non-chemical solid wastes should be placed in the dust bins.

❑❑❑

2

Introduction to Polymers

Chemically polymers are the linear or cross linked long chain molecules having high molecular weight and are built up from simpler low molecular weight compounds, called monomers. The number of monomer units combined to form a polymer is called the degree of polymerization. The combination of similar types or different types of monomers can be used in the manufacturing of polymers. There are many polymers around us in the form of various objects and we are using them in our day to day life comfortably. As these materials have been developed, problems have been raised regarding their disposal. Most of the polymers are not biodegradable and therefore they contribute a significant volume to the garbage. Eco-friendly plastic recycling is an important process for reducing waste and recovering potentially valuable raw materials.

2.1 CLASSIFICATION OF POLYMERS

For convenience, chemists classify polymers in several types, depending on their origin, Composition, Thermal response, Forms and applications, Method of synthesis and Structure.

(i) Classification on the Basis of Origin

Natural and synthetic polymers

Polymers are classified into natural and synthetic polymers. Polymers obtained from the natural sources are called natural polymers and are biodegradable Rubber (from isoprene), cotton, wool, silk, cellulose (from glucose), starch, proteins (from amino acids), leather etc. are the examples of natural polymers. The polymers synthesized in laboratory from the low molecular weight compounds (monomers) are called the synthetic polymers. The various examples of man-made synthetic polymers are Teflon, nylon, polyethylene, polyester, epoxy, vinyl, polyurethane, silicones etc. Most of these polymers are not biodegradable.

(ii) Classification on the Basis of Composition

Organic and Inorganic polymers

On the basis of composition of the backbone chain of the polymers, the polymers are classified as Organic and Inorganic polymers. The polymers whose backbone chain is essentially made up of carbon atoms are called as organic polymers. Examples are Polyethylene, Polystyrene, Polyvinyl alcohol etc. On the other hand the polymers having no carbon atom in their backbone chain are called inorganic polymers. The components of the backbone chain of such inorganic polymers are mainly the atoms like Boron, Nitrogen, Phosphorous, Silicon, Germanium etc. The examples of inorganic polymers are Borazine, Polyphosphonitrile, Polysilicates, Silicon rubber, Glass, Quartz etc. The thermal stability of inorganic polymers is more than that of organic polymers.

Homopolymers and copolymers

On the basis of composition, the polymers are also classified as homopolymers and copolymers. Homopolymers are the polymers formed of only one type of monomer or repeating unit. Examples are polyethylene, Teflon, polystyrene, polyvinyl alcohol etc. Where as co-polymers are the polymers formed by the reaction between two or more different types of monomers or repeating units. Examples are styrene acrylonitrile copolymer (SAN), Styrene isoprene copolymer (SIS), Butadiene Styrene copolymer (BS) etc. Depending on the arrangement of the monomer units, the copolymers are classified as Alternate, Random, Block and Graft copolymers. These copolymers can be diagrammatically represented as shown below:

-A-B-A-B-A-B-A-B-A-B-A-B-A-B-A-B-A-B-A-B- Alternate copolymer

-A-B-B-B-A-A-B-A-A-B-B-B-B-A-B-A-A-B-A-A- Random copolymer

-A-A-A-A-B-B-B-B-A-A-A-A-B-B-B-B-A-A-A-A- Block copolymer

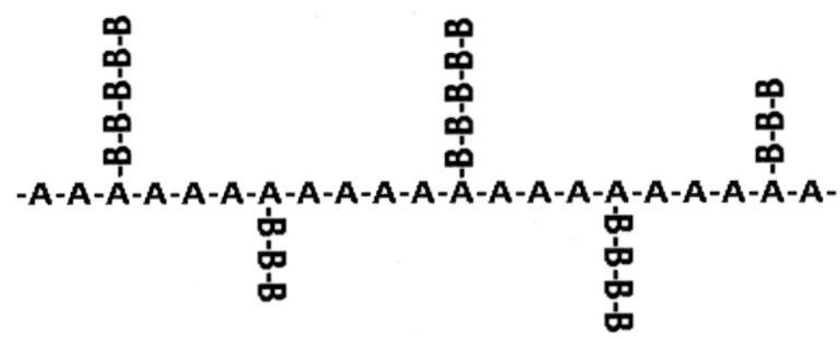

Graft copolymer

(iii) Classification on the Basis of Thermal response

Thermoplastics and Thermosetting polymers

On the basis of the effect of heat or temperature, the polymers are classified as Thermoplastics and Thermosetting polymers. Thermoplastics are the linear or branched polymers which can be repeatedly remolded. That is the process of heating, reshaping and retaining the same on cooling can be repeated several times. The examples of thermoplastics plastics are Polymethyl methacrylate (PMMA), Polyethelene (PE), Polyvinylchloride (PVC), Nylons etc. However the crossed linked polymers which are not recyclable and get converted into an infusible mass on heating are called thermosetting polymers. The examples are Phenolformaldehyde, Vulcanized rubber, Epoxide etc. The thermoplastics are generally soluble in most of the solvents where as the thermosetting polymers are mostly insoluble in all solvent.

(iv) Classification on the Basis of Forms and Applications

Plastics, Elastomers, Fibres and Liquid resins

Depending on the form and the applications the polymers are classified as plastics, elastomers, fibres and liquid resins. Plastics are the materials which can be converted into hard and tough utility articles by the application of heat and pressure. Examples are Polystyrene, Polyvinylchloride etc. Elastomers are the polymers having good strength and elasticity or flexibility. Examples are natural rubber, synthetic rubber, silicon rubber etc. Fibres are the non flexible thread or filament like polymers whose length is at least hundred times more than that of its diameter. Examples of fibres are polyesters, Terylene, Nylons etc. Liquid resin is the polymers which can be used as an adhesives or sealants. Examples are epoxy resins and polysulphides etc.

(v) Classification on the Basis of Method of Synthesis

Addition and Condensation polymers

Based on the method of synthesis the polymers are broadly classified as addition and condensation polymers. Addition polymers are formed by a reaction in which monomer units simply add to one another to form a long-chain (generally linear or branched) polymer. The monomers usually contain carbon-carbon double bonds (whose characteristic reactions are addition reactions). Common examples of addition polymers are polyethylene and Teflon. Condensation polymers are formed by reaction of bi- or polyfunctional molecules, with the elimination of some small molecule (such as water, ammonia, or hydrogen chloride) as a by-product. Common examples of condensation polymers are nylon, Dacron, and polyurethane.

(vi) Classification on the Basis of Structure

Linear, Branched and Cross-linked polymers

On the basis of structure, the polymers are classified as linear, branched and cross-linked polymers. The polymers having long linear backbone chain structure without side branches are called linear polymers. The polymers having side branches attached to the long backbone chain are called branched polymers. Cross-linked polymers are formed when long chains are linked to each other in 3-dimensional structure with tremendous rigidity. Addition and condensation polymers can exist with a cross-linked network, depending on the monomers used in the synthesis. Familiar examples of cross-linked polymers are Bakelite, rubber. These polymers can be diagrammatically represented as shown below:

Linear Polymer

Branched Polymer

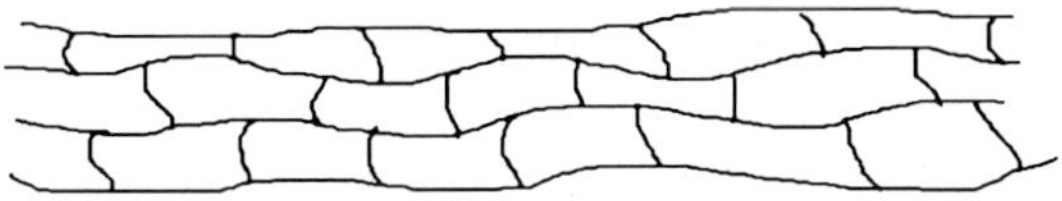

Cross-linked Polymer

(vii) Classification on the Basis of Stereo Regularity

Isotactic, Syndiotactic and Atactic polymers

On the basis of stereo regularity, the polymers are classified as Isotactic, Syndiotactic and Atactic polymers. In Isotactic polymers all side groups (R) are present on the one or same side of the carbon carbon chain. In Syndiotactic polymers all the side groups are present alternatively above and below the plane of the carbon carbon chain. Where as in Atactic polymers the side groups are present randomly on either sides of the chain. These three types of polymers have the same chemical structure but exhibit different chemical properties because of their different configurations and resulting geometrical structure. Atactic polymers are usually have low melting point and easily soluble,

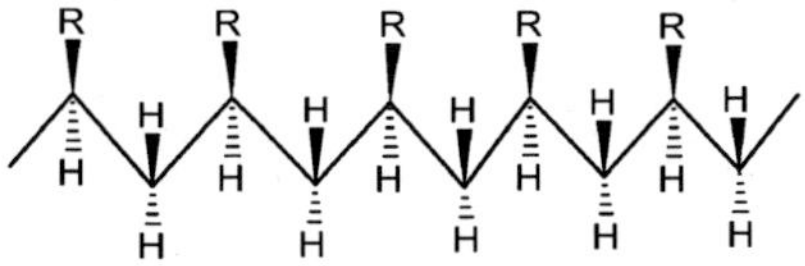

Isostatic polymer

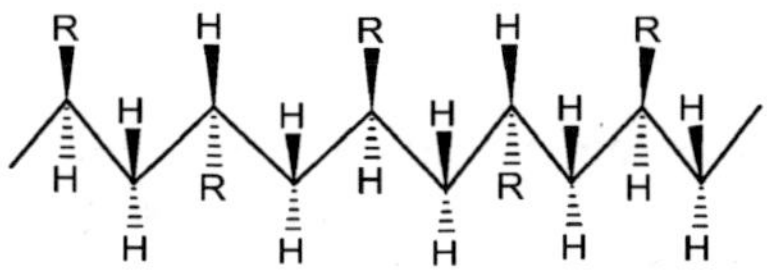

Syndiotactic polymer

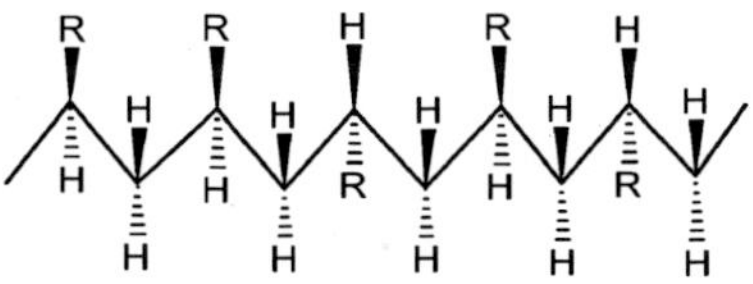

Atactic polymer

while other two polymers have high melting point and less solubility.

2.2 POLYMERIZATION REACTIONS

The process of formation of polymers from low molecular weight compounds i.e. monomers are called the polymerization reactions. Monomers having either unsaturation or reactive functional groups undergo the polymerization reaction. The polymerization reactions are mainly classified as Chain or Addition polymerization and Condensation polymerization reactions.

(i) Chain or Addition Polymerization Reactions

Chain or addition polymerization involves the simple addition of same or different types of monomers having double or tripples bonds i.e.unsaturation. It does not involve the formation of any byproducts. The chain or addition polymerization consists of three major steps, namely initiation, propagation and termination. The process can be carried out either by free radical, ionic or coordination mechanism. The resulting polymers are called addition polymers. The simplest example of addition polymerization is the formation of polythelene from its monomer ethylene as shown below:

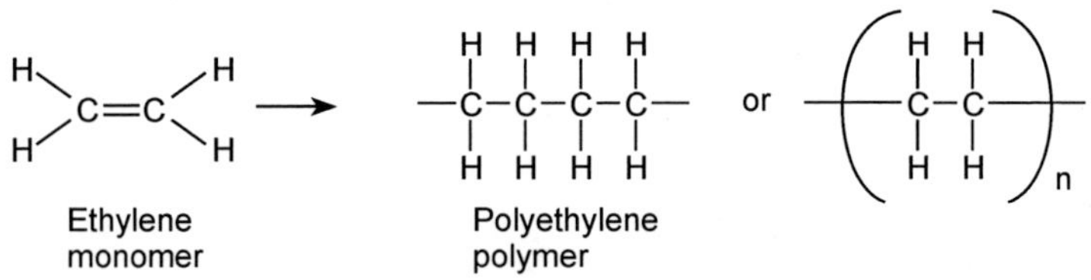

Following table 2.0 shows the example of different addition polymers with monomer units and their applications.

TABLE 2.0. Example of Different Addition Polymer with Monomer Units and their Application

EXAMPLE	MONOMER(S)	POLYMER	USE
Polyethylene	$CH_2=CH_2$	$-CH_2-CH_2-$	Most common and important polymer. Bags, insulation for wires, squeeze bottles
Polypropylene	$CH_2=CH(CH_3)$	$-CH_2-CH(CH_3)-$	Fibers, indoor-outdoor carpets, bottles
Polystyrene	$CH_2=CH(C_6H_5)$	$-CH_2-CH(C_6H_5)-$	Styrofoam, inexpensive household goods, inexpensive molded objects
Polyvinyl chloride (PVC)	$CH_2=CH(Cl)$	$-CH_2-CH(Cl)-$	Synthetic leather, clear bottles, floor covering, phonograph records, water pipe
Polytetrafluoroethylene (Teflon)	$CF_2=CF_2$	$-CF_2-CF_2-$	Nonstick surfaces, chemically resistant films
Polymethyl methacrylate (Lucite, Plexiglas)	$CF_2=C(CO_2CH_3)(CH_3)$	$-CH_2-C(CO_2CH_3)(CH_3)-$	Unbreakable "glass", latex paints
Polyacrylonitrile (Orlon, Acrilan, Creslan)	$CH_2=CH(CN)$	$-CH_2-CH(CN)-$	Fiber used in sweaters, blankets, carpets
Polyvinyl acetate (PVA)	$CH_2=CH(OCCH_3)$, C=O	$-CH_2-CH(OCCH_3)-$, C=O	Adhesives, latex paints, chewing gum, textile coatings
Natural rubber	$CH_2=C(CH_3)CH=CH_2$	$-CH_2-C(CH_3)=CH-CH_2-$	The polymer is cross-linked with sulfur (vulcanization)
Polychloroprene (Neoprene rubber)	$CH_2=C(Cl)CH=CH_2$	$-CH_2-C(Cl)=CH-CH_2-$	Cross-linked with ZnO, resistant to oil, gasoline
Styrene butadiene rubber (SBR)	$CH_2=CH(C_6H_5)$; $CH_2=CHCH=CH_2$	$-CH_2CH(C_6H_5)-CH_2CH=CHCH_2-$	Cross-linked with peroxides. Most common rubber. Used for tires. 25% styrene. 75% butadiene

(ii) Condensation Polymerization Reactions

Condensation polymerization involves the reaction between same or different types of monomers having two or more reactive functional groups like hydroxyl, amino, carboxyl. These polymerization reactions are mostly accompanied by the formation of small molecule as byproduct. The simplest example of condensation polymerization is the formation of polyethylene glycol (PET, Dacron) from its monomers terephthalic acid and ethylene glycol as shown below:

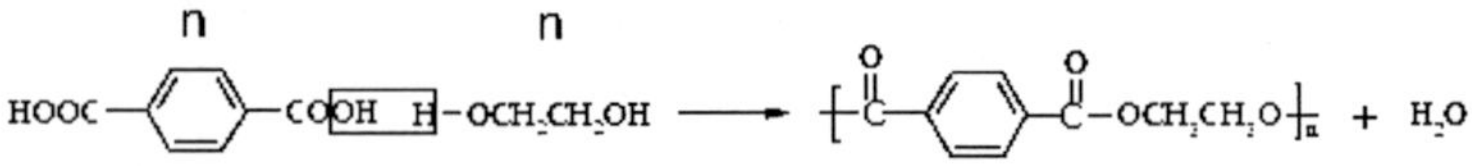

Terephthalic acid   Ethylene glycol   Dacron or PET

Following table 2.1 shows the example of different condensation polymers with monomer units and their applications.

TABLE 2.1. Different Condensation Polymer with Monomer Units and their Application

EXAMPLE	MONOMERS	POLYMER	USE
Polyamides (Nylon)	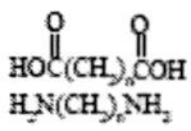	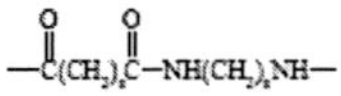	Fibers, molded objects
Polyesters (Dacron, Mylar, Fortrel)	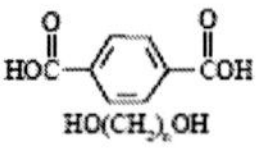	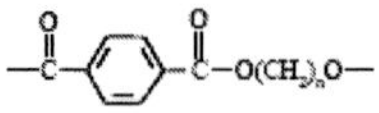	Linear polyesters Fibers, recording tape
Polyesters (Glyptal resin)	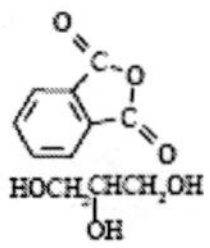	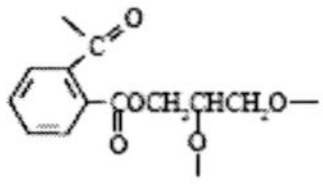	Cross-linked polyester Paints
Polyesters (Casting resin)	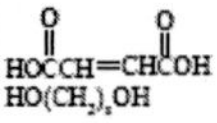	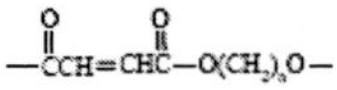	Cross-linked with styrene and peroxide. Fiberglass boat resin
Phenol-formaldehyde resin (Bakelite)	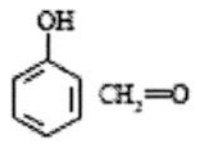	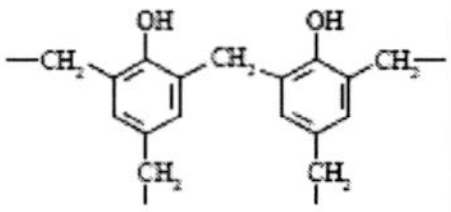	Mixed with fillers. Molded electrical goods, adhesives, laminates, varnishes
Cellulose acetate*	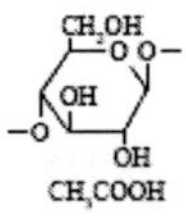	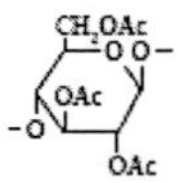	Photographic film
Silicones	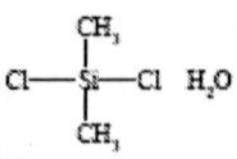	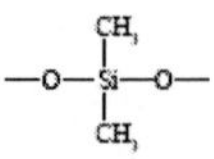	Water-repellent coatings, temperature-resistant fluids and rubbers (CH_3SiCl_3 cross-links in water)
Polyurethanes	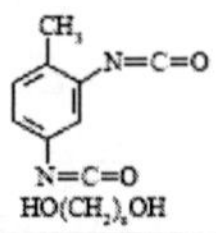	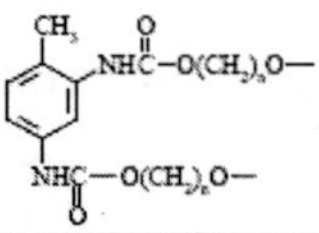	Rigid and flexible foams, fibers

2.3. TECHNIQUES OF POLYMERIZATION

The polymerization can be brought by the various methods or techniques. The criterion for the selection of particular technique depends on the nature of monomer, desired polymerization mechanism and the required physical form of the polymer. Following are most commonly used polymerization techniques.

(i) Bulk or mass polymerization

(ii) Solution polymerization

(iii) Suspension polymerization

(iv) Emulsion polymerization

(v) Interfacial polymerization

(vi) Gas phase polymerization

(vii) Solid phase polymerization

(i) Bulk or mass polymerization

Bulk polymerization is quite simple and carried out in a liquid monomer at definite pressure and temperature. The initiator and the chain transfer agent whenever required to control the molecular weight of polymer are dissolved in the monomer. The whole system is therefore in homogeneous phase. The reaction mass is heated or exposed to a radiation source for the purpose of initiating the polymerization reaction and is kept under agitation for proper heat and mass transfer. If the resulting polymer is soluble in a monomer, the viscosity of the medium increases gradually in the course of polymerization and leads to the formation of monolithic block of polymer. If the polymer is insoluble in monomer, the polymer is obtained as a powder or porous body. The polymer synthesized by this technique is having high purity as no any other additive other than initiator and chain transfer agent can contaminate the polymer. Therefore the polymer obtained can be used as such without isolation and further purification. However an essential disadvantage of this polymerization technique is poor heat exchange in the system. Since polymerization is highly exothermic process, this leads to a considerable over heating of the product, again due to increase in viscosity of the medium during the polymerization process, the diffusibility of the growing polymer chain becomes restricted and the probability of chain termination by collision becomes difficult and rate of polymerization increases enormously. This whole phenomenon is called auto acceleration. The net result is a formation of polymer of non uniform very broad molecular mass distribution. Sometimes the auto acceleration of the uncontrolled exothermic reaction can lead to an explosion. Bulk polymerization can be used to produce polystyrene, polybutadiene, polymethyl methacrylate and other polymers.

(ii) Solution Polymerization

In solution polymerization the reaction medium is usually a suitable inert organic solvent containing dissolved monomer along with the initiator and chain transfer agents whenever used. The two versions are possible.

1. Both monomer and polymer formed are soluble in organic solvent. Polymerization results in a solution of polymer, called lacquer, therefore the method is known as lacquer method. This lacquer may be directly used for coatings or the polymer may be precipitated out of the solution either by the evaporation of the solvent or by adding non-solvent.
2. A monomer is soluble in the organic solvent but the polymer is not. In this case the polymer precipitates out as it is formed and is then separated from the solvent.

The solution polymerization has a disadvantage of resulting the polymers of lower molecular weight due to the chain transfer through solvent molecules if the latter contain mobile atoms or atomic groupings. This gives rise mainly to radicals of low activity which are incapable of initiating a new chain. Collisions of the radicals lead to chain termination, resulting in lower molecular mass of the polymer. The advantages of solution polymerization over bulk polymerization is that the resulting polymer is more uniform in molecular mass due to intensive stirring which improves the heat exchange. Solution polymerization is used in the industrial production of polyacrylonitrile by free radical polymerization and polyisobutylene by cationic polymerization. The block copolymers are made exclusively by this technique.

(iii) Suspension Polymerization

Suspension polymerization usually associated with the polymerization of water insoluble monomers containing dissolved initiators and chain transfer agents whenever used. The monomer remains suspended in water in the form of fine droplets. Thus water acts as a dispersing medium. The suspension is prevented from the aggression of droplets by continuous agitation and stabilized by adding water soluble protective colloids and surface active agents. Here the two separate phases (Aqueous and organic monomer) are present throughout the whole process. The size of the monomer droplets depends on the monomer-water ratio, concentration of the stabilizing agent and the speed of the agitation. Thus each droplet of the monomer can be regarded as the bulk polymerization within a drop. During the polymerization, the immiscible droplet slowly converts from a liquid to a sticky, viscous material and, when reaching a sufficient high molecular weight, forms a solid. Here the water used acts as an efficient heat transfer medium which control the heat evolved during polymerization. Hence the polymer formed is having narrow molecular weight distribution and is in the form of spherical beads or pearls, hence suspension polymerization can also be referred as bead or Pearl polymerization. The isolation of the product formed involves the simple filtration of the beads and removal of the surface active agents and protective colloids by water washing. The water washed dried polymer can be used as such for moulding purpose or as adhesives or for coating by dissolving it in a suitable solvent. Suspension polymerization is used in the production of polystyrene beads, styrene–divinyl benzene copolymer beads and polyvinyl beads etc.

(iv) Emulsion Polymerization

An emulsion polymerization usually associated with the polymerization of slightly water soluble monomers containing dissolved initiators and chain transfer agents whenever used, which form a uniform emulsion in an aqueous dispersion medium i.e. water. The emulsion formed is thermodynamically unstable and hence can be stabilized by adding the emulsifiers. Emulsifiers are the surface active agents that adsorb at the water-monomer interface and form micelles in the surrounding of small amounts of monomer and thus prevent the merging of monomer droplets. The difference between the suspension and emulsion polymerization is that, while both have monomer droplets, the emulsion system has, in addition, micelles in which the monomer is solubilised. If the concentration of the emulsifier added is less than the critical micelle concentration (CMC), the system becomes one of suspension polymerization and only when there is enough emulsifier to form the micelles, we get an emulsion system. Emulsifiers can be anionic (alkali salts of fatty acids and of aryl and alkyl sulfonic acids), cationic (alky amine hydrochlorides and alkyl ammonium halides) or non-ionic (alkyl glycosides and saccharose esters of fatty acids). The polymer emulsions in water resulting from the process are called synthetic latexes. The latexes can be used as such or the polymer can be removed from them by the usual methods of coagulating the colloidal system. Emulsion polymerization process is extensively used by paint, adhesive and coating industries. This is also one of the most important methods for the polymerization of a large number of monomers like vinyl acetate, vinyl chloride, chloroprene, acrylamide, acrylates, and methacrylates. It is also used for the production of various copolymers like acrylonitrile-butadiene-styrene (ABS). Emulsion polymerization has several advantages over other polymerization techniques; for example, it is more rapid than bulk or solution polymerization at the same temperature, the conversion is essentially 100 percent, and the average molecular weight is usually (much) higher than at the same polymerization rate in bulk or solution polymerization. Also, heat dissipation and viscosity control are much less problematic than in bulk polymerization.

(v) Interfacial Polymerization

Interfacial polymerization is a type of step-growth polymerization in which polymerization occurs at an interface of two immiscible liquid phases. One is essentially an aqueous phase containing one monomer and another is an organic phase containing a second monomer. Monomers having highly reactive functional groups like hydroxyl, carboxyl, amine etc. which can react readily at room temperature to form condensation product, are suited for this technique. The product is formed at the interface in the form of thin film while the byproduct dissolves in one of the two liquid phases. The product formed is removed continuously by picking up the thin film using forceps. This leads to the expose of fresh interface resulting in the formation of fresh quantity of polymer. The rate of the polymerization is not high but can be increased by stirring. This greatly increases the interfacial surface area between the two phases. By means of interfacial polycondensation linear polymers with high molecular weight can be obtained. The most common polymer made by this method is polyamide, where diamine dissolved in water and diacid chloride dissolved in organic solvent like carbon tetra chloride or chloroform react to form polyamide and hydrogen chloride byproduct.

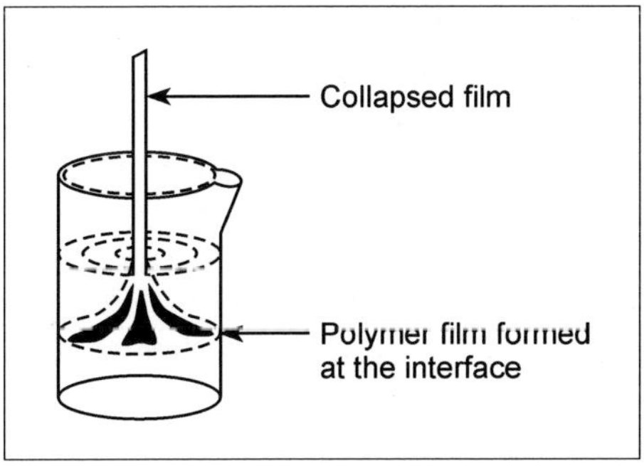

Figure 2.0. Schematic presentation of interfacial polymerization

(vi) Gas or Vapour Phase Polymerization

Gas or Vapour phase polymerization is a widely used polymerization technique for monomers in the gaseous state, such as ethylene (LDPE, HDPE), tetrafluoroethylene (PTFE), and vinyl chloride (PVC). In this polymerization technique polymer formation begins on the inner walls of the reaction vessel and latter on the surface of the particles of already formed polymer. The two methods used in the gas phase polymerization are:

1. Spraying the catalyst (usually Zeigler-Natta type) into the gaseous monomer and

2. Feeding the highly purified monomer gas continuously at particular pressure into the reactor containing fluidized (a finely powdered solid showing the characteristics of a fluid by passing a gas upwards through it) bed column of the catalyst powder (usually Zeigler-Natta type).

In both the cases the polymer is obtained as a free-flowing powder. Beside above many other common thermoplastic polymers can also be made by this method. This includes high volume resins such as polypropylene (PP) polymethyl methacrylate (PMMA), polymethyl acrylate (PMA), polyvinyl acetate (PVA), and polyethylene vinyl acetate (PEVA).

(vii) Solid Phase Polymerization

Many monomers are capable of polymerizing not only in the liquid phase, but also in a crystalline state below their melting point in an inert atmosphere or under vacuum. Since the polymerization is restricted to lower temperatures (below melting point of monomer), thermal activation of the reaction is quite difficult, and hence such polymerization is initiated by irradiating the monomer crystals with X-rays or Gamma rays, fast moving electron beam or other high energy particles. Since the molecular mobility in the solid state is completely restricted, the polymerization occurs at a very slow rate. A good example of solid state polymerization is the preparation of polyformaldehyde by the photo polymerization of solid trioxane monomer.

2.4 SOLUBILITY AND SOLUBILITY PARAMETERS OF POLYMERS

The study of solubility of the polymer is a basic requirement for its characterization and various physical measurements. Polymer dissolution also plays an important role in recycling of the plastics. Unlike non-polymeric materials, polymers do not dissolve instantaneously because of the size differences between polymer and the solvent molecules, the viscosity of the system, and the effects of the texture and molecular weight of the polymer. The dissolution of a polymer into a solvent involves two transport processes, namely solvent diffusion and chain disentanglement. Cross linked polymers do not dissolve but may only swell depending on the extent of cross linking. Hard rubbers or thermosetting resins do not even swell in contact with any solvent. Crystalline polymers do not dissolve except at temperature near their crystalline melting point because crystallinity decreases as the melting point approaches. Branched polymers appear to be more readily soluble than that of their linear polymers. The solubility of a given polymer in various solvents is largely determined by its chemical structure. Polymers will dissolve in solvents having solubility parameters which are not much different from that of the polymer. This principle has become known as *'like dissolves like'*.

The following table 2.2 shows the list of solvents and non solvents of some common polymers.

TABLE 2.2. Solvents and non-solvents of some common polymers

Polymer	Solvents	Non-solvents
High density polyethylene (HDPE)	Above 80°C: Aliphatic, cycloaliphatic and aromatic hydrocarbons, higher aliphatic esters and ketones	All common organic solvents at room temperature
Low density polyethylene (LDPE)	At 40 to 60°C: Aliphatic, cycloaliphatic and aromatic hydrocarbons, higher aliphatic esters and ketones	All common organic solvents at room temperature
Polyethylene oxide	Water , Benzene, Dimethyl formamide (DMF)	Diethyl ether, Aliphatic hydrocarbons
Polystyrene	Benzene, Toluene, Tetrahydrofuran (THF), Dioxane, Ethyl acetate, Butyl acetate, Carbon disulfide, Methyl ethyl ketone	Acetone, Diethyl ether, Alcohols, Saturated hydrocarbons

Polymethyl methacrylate	Benzene, Toluene, Methyl ethyl ketone, Chloroform, Dioxane, Ethyl acetate, Acetic acid, Chlorobenzene	Hexane, Cyclohexane, Methanol, Ethylene glycol, Glycerol
Polyvinyl acetate	Benzene, Toluene, Chloroform, Chlorobenzene, Methanol, Tetrahydrofuran (THF), Dioxane, Acetone, Methyl ethyl ketone, Acetic acid, Dimethyl formamide(DMF), Dimethyl sulfoxide (DMSO)	Saturated hydrocarbons, Cyclohexanol, Diethyl ether, Carbon disulfide, Ethanol
Polyvinyl alcohol	Hot water, Glycerol (hot), Formamide, Dimethyl formamide(DMF), Dimethyl sulfoxide (DMSO) hot	Hydrocarbons, Chlorinated hydrocarbons, Tetrahydrofuran (THF), Dioxane, Esters, Ketones,
Polyvinyl chloride	Tetrahydrofuran (THF), Methyl ethyl ketone, Cyclopentanone, Cyclohexanone, Dimethyl formamide(DMF), Dimethyl sulfoxide (DMSO), Nitrobenzene	Aliphatic and aromatic hydrocarbons, Alcohols, Carbon disulfide, Vinyl chloride
Polymethyl acrylate	Aromatic hydrocarbons, Chlorinated hydrocarbons, Tetrahydrofuran (THF), Esters, Ketones	Aliphatic hydrocarbons, Diethyl ether, Carbon tetrachloride, Methanol, Ethanol
Polyethyl acrylate	Aromatic hydrocarbons, Chlorinated hydrocarbons, Tetrahydrofuran (THF), Esters, Ketones, Methanol, Ethanol	Aliphatic hydrocarbons, Diethyl ether, Cyclohexanol

Polybutyl acrylate	Aromatic hydrocarbons, Chlorinated hydrocarbons, Tetrahydrofuran (THF), Esters, Ketones, Butanol	Aliphatic hydrocarbons, Methanol, Ethanol, Ethyl acetate
Polyacrylic acid	Water, Methanol, Dioxane, Dimethyl formamise (DMF)	Hydrocarbons, Acetone, Diethyl ether
Polyacrylamide	Water	Methanol,
Aliphatic polyesters	Benzene, Chloroform, Formic acid	Aliphatic hydrocarbons, Methanol, Diethyl ether,
Polyethylene terephtalate	Phenol, Chlorophenol, Nitrobenzene	Chlorinated hydrocarbons, Aliphatic alcohols, Ketones, Ethers
Polyurathane	Tetrahydrofuran (THF), Dimethyl formamide(DMF), Formic acid,Ethyl acetate, m-Cresol	Methanol, Diethyl ether, Hydrocarbons
Polyacrylonitrile	Dioxane, Dimethyl sulfoxide (DMSO), Dimethyl formamide(DMF), Sulfuric acid, Nitric acid	Hydrocarbons, Chlorinated hydrocarbons, Alcohols, Diethyl ether, Ketones, Formamides, Acrylonitrile
Nylon 6	Chlorophenol, Formic acid, Acetic acid, Trichloroacetic acid, m-cresol, Sulfuric acid, Phosphoric acid	Hydrocarbons, Alcohols, Chloroform, Ethers, Esters, Ketones
Nylon 66	Phenol, Cresols, Formic acid, Halogenated acetic acid, Sulfuric acid, Phosphoric acid	Hydrocarbons, Aliphatic alcohols, Diethyl ether, Chloroform, Ethers, Ketones
Cellulose triacetate	Chloroform, Dioxane, Acetone, Ethyl acetate, Tetrahydrofuran (THF), Ethyl acetate	Aliphatic hydrocarbons, Benzene, Dichloroethane, Chlorobenzene, Ethanol

Cellulose trinitrate	Acetone, Cyclohexanone, Ethyl acetate, Nitrobenzene, Halogenated hydrocarbons	Aliphatic and aromatic hydrocarbons, Lower alcohols, Diethyl ether, Water
Polybutadiene/ Polyisoprene/ Natural rubber	Aliphatic, Cycloaliphatic, Aromatic hydrocarbons, Chlorinated hydrocarbons, Tetrahydrofuran (THF), Higher ketones, Higher aliphatic esters	Alcohols, Lower ketones and esters, Nitromethane, Water
Polyisobutylene	Aliphatic, Cycloaliphatic, Aromatic hydrocarbons, Chlorinated hydrocarbons	Lower alcohols/ ketones / esters, Nitro methane, Cyclopropionitrile
Poytetrafluoroethylene	Not soluble in any of the strong acids	—
Epoxy resins (cured)	Insoluble in most of the solvents	—
Epoxy resins (uncured)	Alcohol. Dioxane, Ketones, Esters	—

The solubility parameter is an important concept in the theory of solutions and has been shown to be connected to other physical properties such as surface tension, wet ability, the ratio of the coefficient of thermal expansion to compressibility, the boiling points in the case of non-polar liquids, the ultimate strength of materials, and the glass transition temperature of polymers. Therefore, the ability to estimate the solubility parameters can often be a useful tool in predicting systems physical properties and performance.

In 1936, Joel H. Hildebrand proposed a simple definition for a "solubility parameter" that would provide a systemic description of the miscibility behavior of solvents. This solubility parameter is defined as the square root of the cohesive energy density (CED). The cohesive energy density is the amount of energy needed to completely remove unit volume of molecules from their neighbors to infinite separation (an ideal gas). This is equal to the heat of vaporization the compound divided by its molar volume in the condensed phase:

$$\delta = (CED)^{0.5} = \sqrt{\frac{\Delta H_v - RT}{V_m}}$$

This δ is known as the "Hildebrand solubility parameter". If δ_1 and δ_2 are the solubility parameters of solvent and solute respectively. Then miscibility or solubility can be expected or believed to be favorable if δ_1 and δ_2 values are close (*i.e.* $\delta_1 - \delta_2 < 1.8$), but solubility is not favored if the solubility parameter difference is much higher. These predictions are often useful for non polar and slightly polar (Dipole moment < 2 D) systems without hydrogen bonding.

Existence of extensive intramolecular and intermolecular H-bonding in solute or solvent acts as a hindrance to solubility, and in such cases, only solvents, that can favourably engage the solute molecules (or segments of polymer molecules) through solute – solvent H – bonding, can impart or infuse solubility.

Hansen proposed an extension of the Hildebrand parameter method for better understanding of the complex interactions between substances, known as Hansen solubility parameters, have been proposed for polar molecules and hydrogen bonding systems. The basis of his theory lies within dividing total cohesive energy density into three components arising from dispersion forces (E_d), permanent dipole-dipole interactions (E_p) and hydrogen bonds (E_h).

$$CED = E_{coh} = E_d + E_p + E_h$$

If the above equation is divided by molar volume, V_m and then squared, it gives us a new three dimensional approach in which Hildebrand's total solubility parameter squared is equal to the sum of the squares of Hansen's components,

$$\frac{E_{coh}}{V_m} = \frac{E_d}{V_m} + \frac{E_p}{V_m} + \frac{E_h}{V_m}$$

$$\delta^2 = (\delta_d^2 + \delta_p^2 + \delta_h^2)$$

Where δ_d, δ_p, δ_h are Hansen's solubility parameters which represent the three main types of interactions in organic matter and δ is the Hildebrand solubility parameter, $\left(\frac{E_{coh}}{V_m}\right)^{0.5}$.

Thus in Hansen's approach the Hildebrand solubility parameter is split into three components; polar, dispersion, and hydrogen bonding. Hence, can also be referred as 3D solubility parameters. The three components are empirically adjusted to define the miscibility characteristics of the solvent. Solvents with similar Hansen solubility miscible in all proportions where as dissimilar values yield limited solubility.

Hildebrand and Hansen solubility parameters are useful for selecting solvents and additives in formulations, for the blending of polymers, for the control of kinetics and monomer sequence distributions in copolymers, and for the proper selection of time release formulations in the delivery of pharmaceuticals.

The conventional units for the solubility parameter are (calories / $cm^3)^{1/2}$, or $cal^{1/2}\ cm^{-3/2}$. The SI units are $J^{1/2}\ m^{-3/2}$, equivalent to the $(Pascal)^{1/2}$. 1 calorie is equal to 4.184 J, 1 $cal^{1/2}$ $cm^{-3/2} = (4.184\ J)^{1/2}\ (0.01\ m)^{-3/2} = 2.045\ mPa^{1/2}$.

The experimental method of solubility parameters determination involves the study of intrinsic viscosity of polymer in different solvents at constant temperature. The solubility parameter of the polymer will be equal to the solubility parameter of that solvent for which the intrinsic viscosity [η] is maximum.

Typical values of Hildebrand solubility parameters for some monomers solvents and polymers are listed in following table.

TABLE 2.3. Solubility Parameter values of some Solvents and Polymers

Solvent	δ_1 (Cal/ $cm^3)^{1/2}$	Polymer	δ_2 (Cal/ $cm^3)^{1/2}$
Methanol	14.5	Cellulose	15.65
Ethanol	12.7	Cellulose triacetate	13.60
2-Ethyl hexanol	9.5	Poly (vinyl alcohol)	12.60
Chloroform	9.3	Poly (vinyl chloride)	9.42
n-Butyl acetate	8.5	Poly (vinyl acetate)	9.59
Dibutyl amine	8.1	Polybutadiene	8.4
Difluorodichloromethane	5.1	Poly (methylmethacrylate)	9.45
Acetonitrile	11.8	Polyacrylonitrile	12.50
1,4-Dioxane	10.0	Polytetrafluoroethylene	6.2

Acetone	9.9	Poly(methyl cyanoacrylate)	14.00
Benzene	9.2		
Toluene	8.9	Poly (dimethyl siloxane)	7.35
Carbon tetra chloride	8.6	Polyisobutene	7.85
		Benzyl cellulose	12.33
Ammonia	16.3	Polystyrene	9.10
Cyclohexane	8.2	Nylon 66	13.6
n-Decane	6.6	Polyethylene	7.70
Glycerol	16.5	Polypropylene	9.3
Formamide	19.2	Natural rubber	7.90
Acetic acid	10.1	Epoxy resin	10.91
Acrylic acid	12.0	Polymethacrylonitrile	10.71
Pyridine	10.7	Poly (ethylene terephthalate)	10.70
Xylene	8.8		
DMSO	12.0	Polyethyleneoxide	9.9
Diethylether	7.4	Polyoxymethylene	9.9
Water	23.4	Polyethyl methacrylate	9.0
		Polyisoprene (Natural rubber)	8.2

2.5 GLASS TRANSITION TEMPERATURE (T_g)

The knowledge of T_g is essential in the selection of materials for various applications. The study of effect of temperature on physical state of polymer can be used to understand the glass transition and melting point temperatures of polymer. The temperature below which a polymer is hard or brittle and above which it is soft or flexible is called as the glass transition temperature. The hard or brittle state is known as glassy state and soft or flexible state is known as rubbery state. One of the simplest ways to define the transition temperatures T_g and melting point temperature Tm is to consider the changes in specific volume that take place on heating. The Volume-temperature curves can be determined experimentally using dilatometry, although this is not the preferred method of determining transition temperatures.

For low molecular weight crystalline materials the transition from the solid to the liquid state is sharp and occurs at a well defined temperature known as melting point temperature (T_m). The following diagram Fig. 2.1 shows a plot of specific volume versus temperature for a fully crystalline solid. The transition to the liquid phase is accompanied by a discontinuous change in volume at T_m (and similarly in the enthalpy). This discontinuity is characteristic of a first order transition. (The slopes of the volume-temperature curve below T_m and above Tm are related to the thermal expansion coefficients of the solid and liquid phases, respectively).

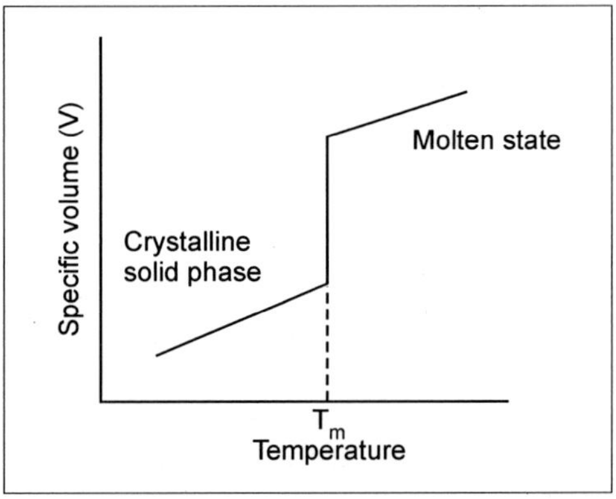

Fig. 2.1. A plot of specific volume versus temperature for a crystalline low molecular weight solid.

The situation differs considerably for a polymer. The nature of the curve showing specific volume versus temperature for fully crystalline, amorphous and partially crystalline polymer are given below Fig. 2.2. In case of crystalline polymers the melting process does not occurs at sharp temperature but over a finite temperature range. This is shown by a gradual bend in the plot just before T_m (curve a). This range of temperature is dependent on the polydispersity of the sample. Broader the molecular weight distribution, wider the range. In case of amorphous polymers there is slight change in the slope of the curve near the glass transition and then there is all of sudden increase in the slope steeply. It can be seen from the graph that beyond T_g, there is no change in the slope even at melting point temperature (curve b). This indicate that there is no appreciably change in the specific volume while moving from rubbery to molten state, *i.e.* the amorphous polymer does not undergo true melting.

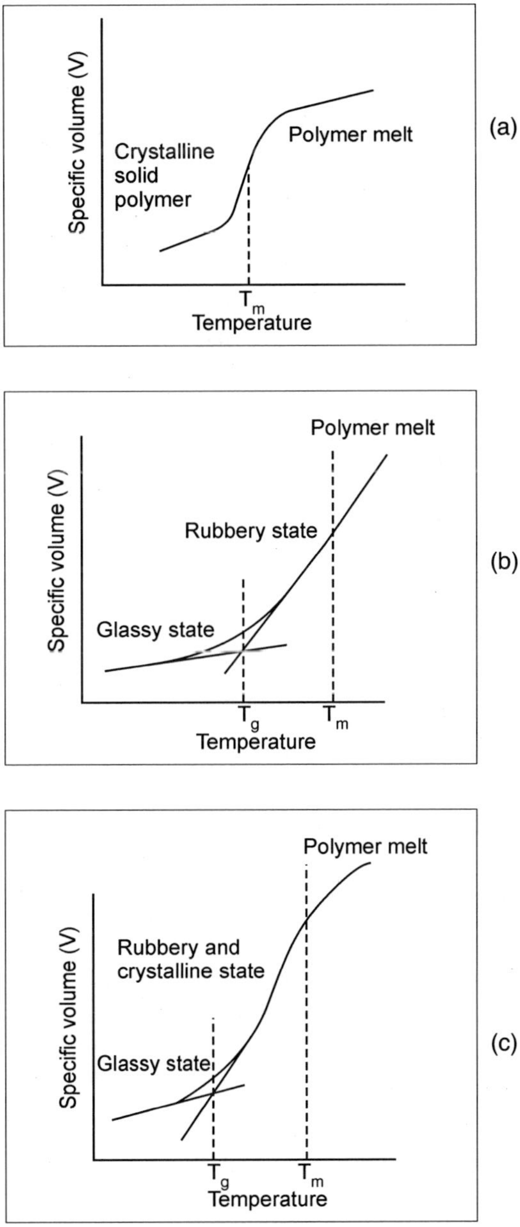

Fig. 2.2. The nature of specific volume -Temperature curve for (a) fully crystalline (b) amorphous and (c) partially crystalline polymer.

However in case of partially crystalline polymer (curve c), mixed behavior is observed, there is change in the slope of the curve at two different temperature ranges, the lower temperature range represent the glass transition where as the higher temperature represents the melting region.

Thus the glass transition does not involve any discontinuity in the V-T curve (and similarly in the enthalpy versus T curve) is often called a second order transition.

The glass transition temperature of a polymer depends on the chain geometry, chain flexibility and molecular aggregates. Following table shows the glass transition temperature (T_g) of some common polymers.

TABLE 2.4. Showing the Glass Transition Temperature (T_g) of the Polymers

Sr. No.	Name of Polymer	Tg (°C)	Tm (°C)
1.	Poly vinyl Pyrrolidone	168	
2.	Cellulose triacetate	157	
3.	Polytetrafluoroethylene	125	325
4.	Polyethylene (HDPE)	125	135
5.	Polyacrylic acid	106	
6.	Polymethylmethacrylate	105	
7.	Polystyrene	100	240
8.	Polyacrylonitrile	97	
9.	Polyvinyl Chloride	87	227
10.	Polyvinyl Alcohol	85	220
11.	Polyethyleneterephthalate	70	265
12.	Polyethylmethacrylate	65	
13.	Nylon 6	52	225
14.	Nylon 66	50	265
15.	Cellulose acetate butyrate	50	
16.	Polyvinyl Acetate	32	
17.	Polymethylacrylate	06	
18.	Polyethylacrylate	-24	
19.	Polyethylene adipate	-50	

2.6. MOLECULAR WEIGHT OF POLYMERS

Unlike simpler pure compounds, the Polymers consist of repeating units called monomers chemically bonded to one another into long chains. In understanding the physical properties of a polymer (such as density, mechanical strength, solubility and brittleness) requires knowledge of the length of the polymer chains. The polymer molecules produced by polymerization reaction have chain lengths that are governed by the polymerization mechanism and by the condition prevailing during the polymerization process. Also, the polymerization process, whether proceeding by simple addition or condensation result in a mixture of polymer chains of unequal length. Chain length is often expressed in terms of the molecular weight of the polymer chain, related to the relative molecular mass of the monomers and the number of monomers connected in the chain. Therefore the molecular weight of polymer is not a single value and can be described as some average molecular weight calculated from the molecular weights of all the chains in the sample. Thus molecular weight can be expressed in terms of either number average molecular weight or weight average molecular weight.

(i) Number average Molecular Weight (M_n)

The number average molecular weight is the statistical average molecular weight of all the polymer chains in the sample. It can be calculated by multiplying the molecular mass of the each chain by the number of molecular chains having that mass and the resulting products are added up, giving the sum of molecular masses of all the molecules, then this sum is divided by the total number of molecules i.e. the total number of chains, as given by the following mathematical equation.

$$M_n = \frac{N_1 M_1 + N_2 M_2 + N_3 M_3 + N_4 M_4 + \ ......+ N_n M_n}{N_1 + N_2 + N_3 + N_4 + \ ......+ N_n}$$

$$M_n = \frac{\Sigma N_i M_i}{\Sigma N_i}$$

Where M_i is the molecular weight of a chain and

N_i is the number of chains of that molecular weight.

(ii) Weight average molecular weight (M_w)

The weight average molecular weight is the sum of the product of contribution of molecular weight of each molecule or chain to the total molecular mass of the polymer and the molecular weight of that polymer molecule or chain.

The total molecular mass of the polymer is equal to $\Sigma N_n M_n$, and hence the contribution of particular polymer chain or polymer molecule (X_i) to the total molecular mass of the polymer is equal to,

$$X_i = \frac{N_i M_i}{N_1 M_1 + N_2 M_2 + N_3 M_3 + N_4 M_4 + \ldots\ldots + N_n M_n} = \frac{N_i M_i}{\Sigma N_n M_n}$$

Therefore the weight average molecular weight (Mw) is given by the equation,

$$M_w = X_1 M_1 + X_2 M_2 + X_3 M_3 + X_4 M_4 + \ldots\ldots + X_n M_n = \Sigma X_n M_n$$

$$M_w = \frac{N_1 M_1{}^* M_1}{\Sigma N_n M_n} + \frac{N_2 M_2{}^* M_2}{\Sigma N_n M_n} + \frac{N_3 M_3{}^* M_3}{\Sigma N_n M_n} + \ldots + \frac{N_n M_n{}^* M_n}{\Sigma N_n M_n}$$

$$M_w = \frac{N_1 M_1^2 + N_2 M_2^2 + N_3 M_3^2 + \ldots + N_n M_n^2}{\Sigma N_n M_n}$$

$$M_w = \frac{\Sigma N_i M_i^2}{\Sigma N_i M_i}$$

Where M_i is the molecular weight of a polymer molecule or chain and

N_i is the number of chains having that molecular weight.

2.7 CONDUCTING POLYMERS

In the beginning of discovery of polymers, they were considered as a no conducting insulating materials. But in the year 1977 the scientists Alan G. MacDiarmid, Alan J. Heeger, and Hideki Shirakawa at the University of Pennsylvania demonstrate that the polyacetylene which is an insulator becomes electrical conductor on treatment with strong oxidizing or reducing agent.

For their discovery and development of conducting polymers they received the Nobel Prize in chemistry in the year 2000. Since then till now the varieties of conducting polymers as shown below have been introduced based on monomers like pyrroles, thiophenes, anilines or aromatic hydrocarbons. The important feature of all these conducting polymers is having alternate π and σ bonds along the polymer backbone resulting in a π-conjugation. The electronic structure and the electronic properties of conducting polymers can be explained by the molecular orbital theory.

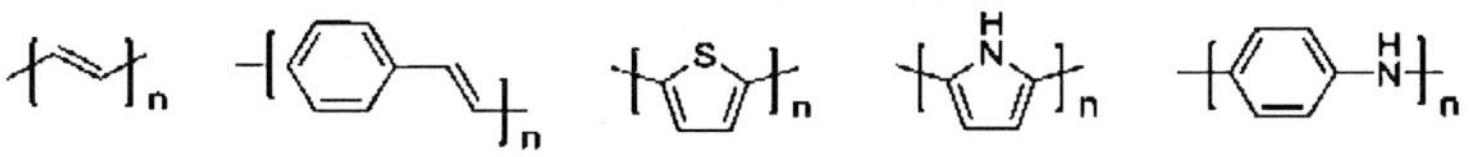

Polyacetylene Polyparaphenyl-envinylene Polythiophene Polyphrrole Polyaniline

The normal insulating polymers can be made conducting by incorporating conductive fillers or additives into them. The examples are given below,

Rubber + Graphite powder → conducting rubber

Poly vinyl chloride powder + Nickel powder → conducting PVC

Poly ester films + Aluminum powder → conducting metalized Poly ester films

The electrical conductivity of polymers depends on the factors like the method of synthesis, the method of processing, crystallinity of polymers and temperature. The conductivity increases with the increase in degree of crystallinity of polymers. Initially the conductivity increases slowly up to 50 % crystallinity and then it increases very rapidly.

❑❑❑

3

Characterization of Polymers

Characterization of polymers is important for, when new materials are synthesized, when a competitive product is being evaluated, or when the performance of a product needs to be improved. The polymers synthesized in laboratory by different polymerization techniques after purification are need to be characterized by various chemical, physical, analytical, thermal and spectral analysis for determination of their molar mass, solubility, melting point, glass transition temperature, thermal stability and structure.

3.1. DETERMINATION OF MOLECULAR MASS OF POLYMERS

A wide range of chemical, physical methods have been employed for the determination of the molecular mass of the polymers. The chemical method includes end group analysis based on halogenation, saponification, acetylation etc. of reactive functional end groups. Physical methods include Viscometry, Osmometry etc. methods. These methods are discussed below.

3.1.1. End Group Analysis

The number average molecular weight of polymer may be determined experimentally by measuring the total number of moles of polymer in a known weight of sample. The method of the analysis for the determination of number of moles of polymer, must respond equally to each molecule without regard to its chain length. For a linear polymer, every molecule has two end groups, which may be identical, as in polyethylene glycol (PEG), or different, as in Polyethylene terephthalate (Dacron). Thus, the end groups can be analyzed by titration of the reactive functional group with suitable reagent and the number of polymer molecules may be calculated using simple stoichiometric relationships. This method for determining Mn is called end group analysis. Linear condensation polymers such as polyesters and polyamides are most suited for end group analysis. The end group analysis method is used for the polymers with molecular weight less than 50,000. The variety of functional end groups and their methods of analysis are given below.

Hydroxyl Group

In the determination of hydroxyl groups (phenolic or alcoholic), a accurately weighed polymer sample is acetylated with excess acetic anhydride, and liberated acetic acid is titrated with standardized base to the end point with phenolphthalein indicator.

Epoxy (Oxiranc) Group

Chemical method used for the determination of epoxy groups in the polymer is based on reaction with hydrobromic acid.

Carbonyl Group

One of the common chemical methods for determining carbonyl compounds consists of converting them into hydrazones by using hydrazine hydrochloride or phenyl hydrazine reagent.

Carboxyl Group

In the determination of carboxyl group, accurately weighed polymer sample is dissolved in an appropriate solvent such as acetone and titrated with standard base to a phenolphthalein end point.

Amine Group

Chemical method used for the determination of amine group in polymer is based on reaction with hydrobromic acid.

3.1.2. Viscometry Method

The viscosity of a liquid is the property of showing resistance to flow. In the case of solutions, the viscosity of the solvent can alter significantly depending on the type and concentration of the solute. In case of polymer solution, due to their large size and high molecular weight, the viscosity of the solutions changes considerably with concentration. The number average molecular mass of polymer can be obtained by measuring the viscosity of different solutions of known polymer concentration along with solvent and using the following equations.

1. Relative viscosity (η_r): $\eta_r = \dfrac{\eta}{\eta_0}$

   Where η is the viscosity of polymer solution

   η_0 is the viscosity of pure solvent

2. Specific viscosity (η_{sp}): $\eta_{sp} = (\eta_r - 1) = \dfrac{\eta}{\eta_0} - 1$

3. Reduced Viscosity ($\eta_{red.}$): $\eta_{red} = \left(\dfrac{\eta_{sp}}{C}\right)$

   Where C is the concentration of polymer solution in %

4. Inherent viscosity ($\eta_{inh.}$): Inherent viscosity also called logarithmic viscosity number,

$$\eta_{inh} = \frac{\ln \eta_r}{C}$$

5. Intrinsic viscosity [η]: Plot the graph between reduced viscosity $\left(\dfrac{\eta_{sp}}{C}\right)$ **or** inherent viscosity $\left(\dfrac{\ln \eta_r}{C}\right)$ along Y-axis and % concentration along X-axis. Determine intrinsic viscosity [η] of the polymer solution by extrapolating the straight line obtained to zero concentration.

$$[\eta] = \lim_{c\to 0}\left(\frac{\eta_{sp}}{C}\right) = \lim_{c\to 0}\left(\frac{\ln \eta_r}{C}\right)$$

6. Average molecular weight of the polymer (M) is related with molar mass of polymer by the Mark Houwink's equation given by.

$$[\eta] = K\,M^{\alpha}$$

Where M is the average molecular weight of polymer.

[η] is the intrinsic viscosity of the polymer solution.

K & α are Mark Houwink's constants for given polymer and solvent pair at the temperature of viscosity measurement.

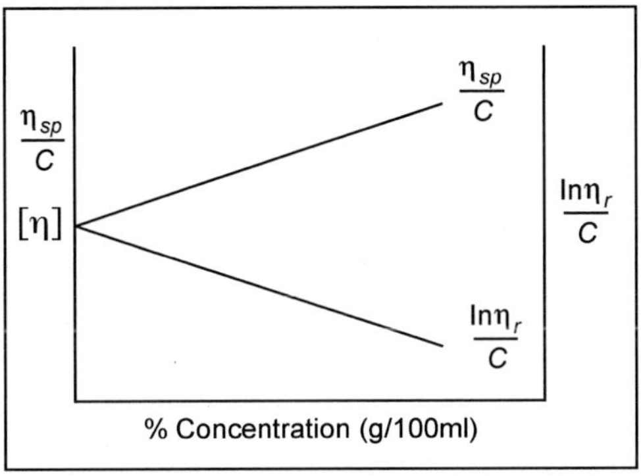

Fig. 3.0. Plot of reduced viscosity $\left(\frac{\eta_{sp}}{C}\right)$ or inherent viscosity $\left(\frac{\ln\eta_r}{C}\right)$ and % concentration to determine intrinsic viscosity [η] of the polymer.

3.1.3 Osmometry Method

Osmometry method for determining molecular mass of polymer based on the colligative property (osmotic pressure) which depends on the number of dissolved solute molecules in a solution. When pure solvent is separated from a solution by a semi-permeable membrane that allows passage of solvent but not solute molecules, solvent will flow through the membrane into the solution. As the liquid level rises in the solution compartment, the hydrostatic pressure increases until it prevents further passage of solvent or, more exactly, until solvent flow is equal in both directions. The pressure at equilibrium is the osmotic pressure. A schematic representation of an osmometer is given in following Fig. 3.1.

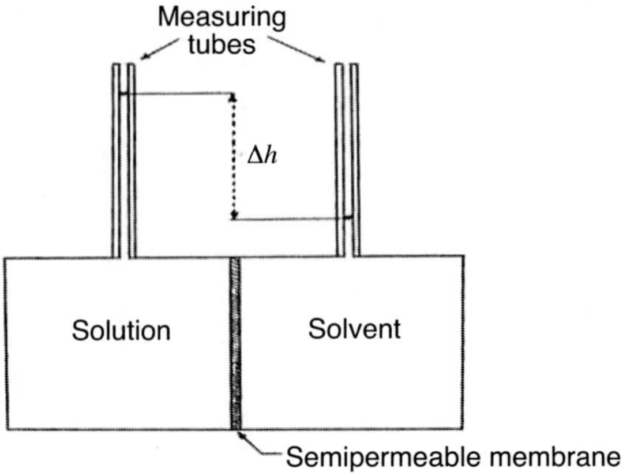

FIG. 3.1. Schematic representation of a membrane osmometer.

Osmotic pressure (π) is related to molecular weight by the van't Hoff equation

$$(\pi \mid c)_{c\to 0} = \frac{RT}{Mn}$$

Where (π/c) = Reduced osmotic pressure of solution

$(\pi/c)_{C=0}$ = Limiting osmotic pressure of solution

C = Concentration of polymer solution in g/100 ml

R = Gas constant, 0.8205 Lit. atm mol^{-1} K^{-1}.

Mn = The number average molecular weight of polymer solution

T = Temperature of measurement in Kelvin

Thus the plot of reduced osmotic pressure (π/c in atm.) along Y – axis against the percentage concentration (g/100ml) along X-axis and Extrapolating the straight line obtained to zero concentration i.e. $(\pi/c)_{C=0}$ as shown in the following Fig. 3.2 can be use to determine the Number average molecular weight of the polymer.

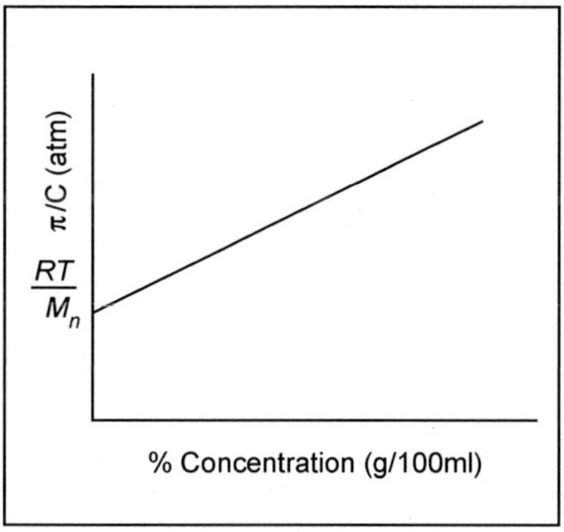

FIG. 3.2. Plot of reduced osmotic pressure (π/c in atm.) versus concentration (g/100ml).

3.1.4 Gel Permeation Chromatography/Size Exclusion Chromatography (GPC/SEC)

GPC/SEC is a type of liquid chromatography technique consisting of a stagnant liquid present in the pores of gel beads column as the stationary phase and a flowing liquid through the column i.e. eluent as the mobile phase. The mobile phase can therefore flow between the beads and also in and out of the pores in the beads. The separation mechanism based on the size of the polymer molecules in solution rather than any chemical interactions between polymer molecules and the stationary phase hence GPC is also referred as size exclusion chromatography (SEC). This technique gives information regarding the size of polymer molecules in solution that can be used for fractionation and to determine the molecular weight of the polymer through the use of a calibration curve.

The basic components of GPC/SEC instrument are,

1. A column: It holds the stationary phase.
2. A pump: It pushes the solvent through the column of the instrument.
3. An injection port: It introduces the test sample into the column.

4. Detector: It detects the components as they leave the column.
5. Computerized controller and analyzer: It controls the different parts of the instrument, analyzed the result and displays the results using software.

The polymer sample is first dissolved in a suitable solvent. In the solution, the polymer molecules are present in the form of spherical coils. The size of the spherical coils depends on the molecular weight i.e. the chain length of the sample. Higher the molecular weight, larger will be the size of spheres. These coiled up polymer molecules solution is then introduced into the mobile phase and flow into the GPC/SEC column. The dissolved polymer molecules move through the column along with the liquid mobile phase. As the polymer coils move down the column, the polymer coils of size larger than that of the biggest pores in the beads, do not enter the pores and so are carried straight down the column by the mobile phase. The polymer coils of size smaller than that of the biggest pores in the beads can enter the larger, but not the smaller pores thus occupy some but not all of the available stationary phase pores. The polymer coils that are smaller than the smallest pores in the beads can enter any of the pores and so can potentially occupy all of the stationary phase pores. The process of diffusion into and back out of polymer coils from pores occurs repeatedly as they travel down the column. As a result, small polymer coils that can enter many pores in the beads take a long time to pass through the column and therefore exit the column slowly. Conversely, large polymer coils that cannot enter the pores take less time to leave the column, and polymer coils of intermediate size exit the column somewhat between these two. Thus, the way in which the samples elute from the column depends very much on the size of the pores in the beads.

As the components exit the column they are detected by the detector, and the elution behavior of the sample is displayed as a graph called chromatogram. This chromatogram shows how much material emerged out of the column at any instance, with respect to their molecular weight. This is the primary separation according to elution volume. Which is then converted to time for ease of measurement, on the assumption of a constant flow rate.

The time required for a group of molecules of the same size (a fraction) to emerge from the column is called the retention time.

The data produced by the chromatogram is then compared with a calibration curve that shows the elution behavior of a series of polymers for which the molecular weights are known, to generate molecular weights and molecular weight averages of the unknown polymer. This allows studying the molecular weight distribution of the sample. The commonly calculated average is actually called the number average molecular weight.

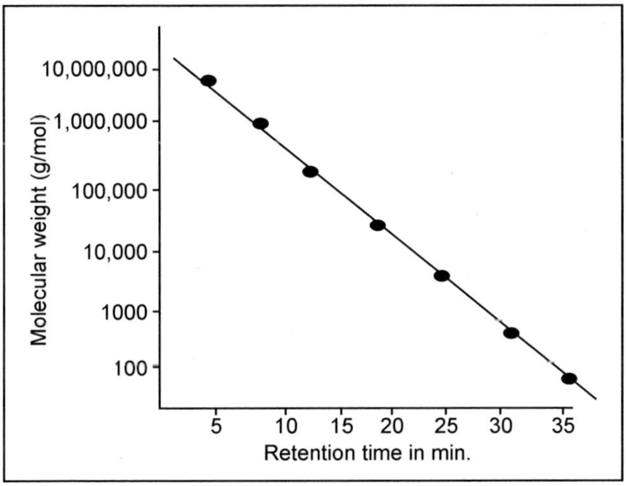

FIG. 3.3. A calibration graph used to determine the molecular weight of a polymer from its retention time.

3.2 THERMAL ANALYSIS

Thermal analysis of polymers is useful for the study of thermal behaviour of polymers. The various thermal analysis of polymers includes thermogravimetric analysis (TGA), differential thermal analysis (DTA) and differential scanning calorimetry (DSC).

3.2.1 Thermogravimetric Analysis (TGA)

Thermogravimetric analysis is most useful for the thermal analysis of polymeric materials especially thermoplastics, thermosets, elastomers, composites, blends, films, fibers, coatings and paints. It is based on the measurement of amount and rate of change in the mass of a polymer sample as a function of temperature or time in air, nitrogen, oxygen, or specially controlled atmosphere resulting a graph of mass versus temperature. The measurements are used primarily to determine the thermal stability, oxidative stability, moisture content, volatility, ash content, effects of reactive atmospheres on materials, percentage of fillers, degradation temperature and decomposition kinetics. The compositional characterization information can also be study through the use of Auto Stepwise TGA, where the TGA instrument automatically heats the sample and then holds it under isothermal conditions when the instrument detects a significant weight loss. This provides the highest possible separation of overlapping decomposition events and provides for the most accurate compositional analysis of polymers. Thus TGA measurements provide valuable information that can be used to select materials for certain end-use applications, predict product performance and improve product quality.

A typical TGA instrument consists of three major components:

1. Sensitive recording balance
2. Furnace and associated controller/atmosphere management
3. A data analyzer including recorder, plotter and any other peripheral equipment.

A sample for thermo gravimetric analysis must be in a solid or a liquid form. The sample is confined within an inert sample pan. Usual sample sizes for commercial instruments are in the range of 0.1 mg to 10 g with heating rates of 0.1 to 50 °C/min. The TGA operates from ambient temperature to an upper limit of 1000 °C with the most commonly employed constant heating rates 10, 15, 20, 25 and 30 °C per minute. The analyzer can detect weight changes of approximately 1 millionth of a gram. Following Fig. 3.4 shows a series of thermo gravimetric curves for common polymers in nitrogen atmosphere.

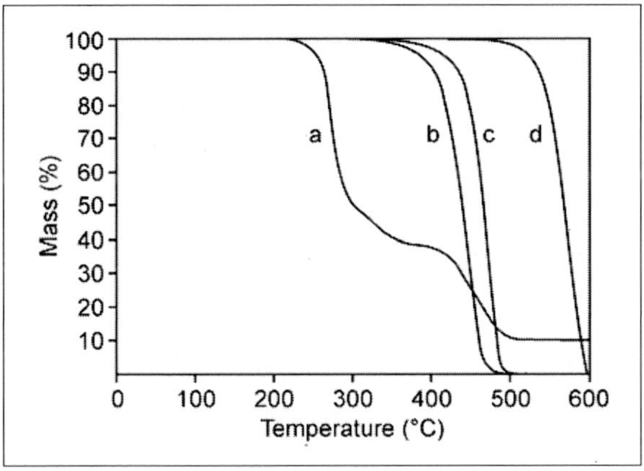

FIG. 3.4. TG curves for polymers in nitrogen heated at 5 °C/min (a) polyvinyl chloride (b) nylon-6 (c) low-density polyethylene (d) polytetrafluoroethylene.

The initial small weight loss up to 100°C is due to the removal of volatile matters and water from the sample which are helpful to determine the volatility and moisture content of the polymer. Sudden loss of weight at a particular temperature gives the decomposition temperature of the polymer.

TGA decomposition information can be used to predict the lifetimes of some polymeric materials, such as the coatings for electrical or telecommunication cables. The sample is heated at three or more different heating rates. The use of the different heating rates, changes the time scale of the decomposition event. The faster the applied heating rate, the higher the given decomposition temperature. The following Fig. 3.5 shows TG curves for low density polyethylene (LDPE) at different heating rates. This approach establishes a link between time and temperature for the polymer decomposition and this information can be used to model the decomposition kinetics. The analysis of shape of curve reveals the kinetics of decomposition, activation energy and frequency factor in Arrhenius equation.

The filler content of the polymer corresponds to the percentage residue left after the complete decomposition of the polymer. This is due to the fact that the fillers do not decompose under the experimental condition while the plastic undergo the

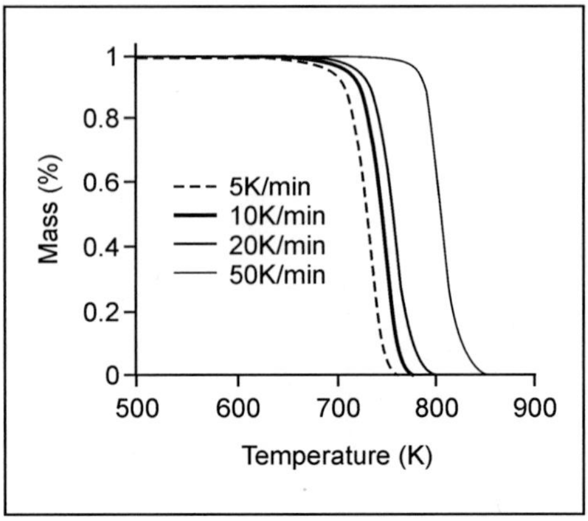

FIG. 3.5. TG curves for low density polyethylene (LDPE) at different heating rates

complete decomposition. The TGA curve (a) for polyvinyl-chloride in the above Fig. 3.4 shows that, it does not undergo complete decomposition. The percentage mass or residue left gives the filler content of the PVC.

3.2.2. Differential Thermal Analysis (DTA)

In differential thermal analysis the sample and the inert reference substance (glass or alumina), undergoing no thermal transitions in the temperature range of interest are heated at the same rate. The temperature difference between the sample and the reference is measured and plotted as a function of sample temperature. The temperature difference is finite only when the heat is being evolved or absorbed because of exothermic or endothermic activity in the sample. This technique is not quantitative in terms of determining energy changes although it is capable of detecting the occurrence and energetic nature of physical and chemical transformations like melting, vaporization, chemical reaction, crystal transition and glass transition. When a polymer is melted or crystallized, heat is absorbed or liberated and these effects are observed in the form of endothermic and exothermal peaks in DTA curves.

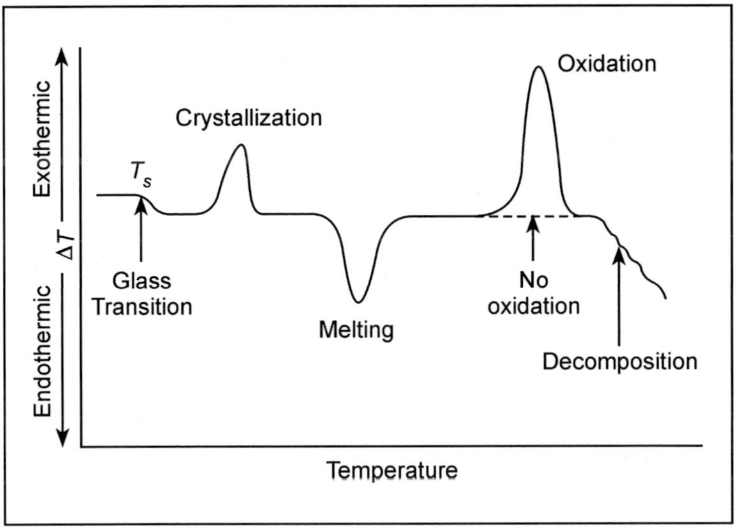

FIG. 3.6. Typical DTA curve showing the occurrence of different physical and chemical transitions in polymer

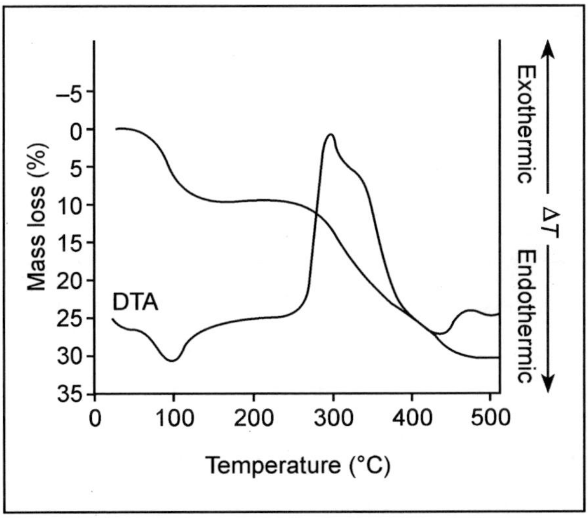

FIG. 3.7. Typical TGA & DTA curve showing the occurrence of different physical and chemical transitions in polymer

3.2.3. Differential Scanning Calorimetry (DSC)

Differential Scanning Calorimetry (DSC) is a thermal instrumental analytical technique in which the rate of absorption of heat by a polymer is compared with that of a standard such as glass or alumina by maintaining a thermal balance between the reference and sample by changing a current passing through the heaters under the two chambers. The circuitry is programmed to give a constant temperature for both the reference and the sample compartments. The heating of a sample and reference proceeds at a predetermined rate until heat is emitted or consumed due to certain physical or chemical transition of the sample. If an endothermic occurrence takes place, the temperature of the sample will be less than that of the reference. Excess current is fed into the sample compartment to raise the temperature to that of the reference. The current necessary to maintain a constant temperature between the sample and reference is recorded. The plot of this power difference against temperature gives the DSC thermogram. The area under the resulting curve is a direct measure of the heat of transition. For DSC analysis sample size range from 0.5 mg to 10 g is required. Typical DSC curve showing the occurrence of different physical transitions in polymer is given below in fig. 3.8.

The following measurements can be made from the DSC analysis

- Glass transition temperature
- Melting point
- Latent heat of melting
- Latent heat of crystallization
- Endothermic and exothermic natures of transitions
- Degree of Crystallinity
- Phase changes
- Specific heat capacity

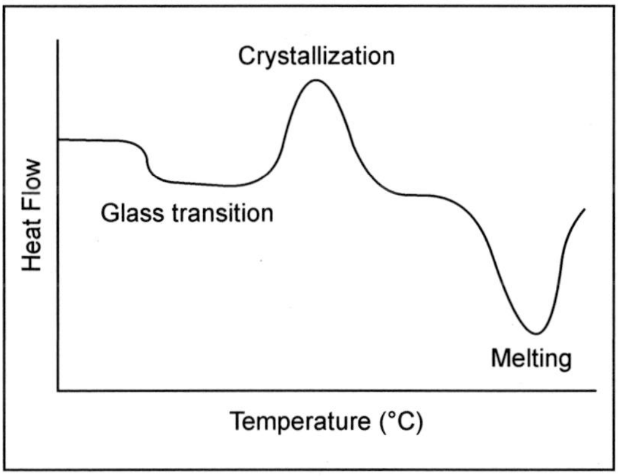

FIG. 3.8. Typical DSC curve showing the occurrence of different physical transitions in polymer

3.3. SPECTRAL CHARACTERIZATION

Spectroscopy is an important tool for the characterization of polymers. This technique is based on the interaction of different electromagnetic radiations with matter resulting in a spectrum. The interpretation of the spectrum is helpful in the characterization of polymers in terms of its structure, functional groups, molecular weight and crystallinity determination.

3.3.1. Infrared Characterization

Infrared spectroscopy is an extremely useful technique to identify polymer structure. When the frequency of incident IR radiation is equal to the frequency of a specific molecular vibration, the molecule absorbs the radiation which appears as an absorption peak in the IR spectra. The IR spectra interpretation involves the study of the position of characteristic absorption peaks and their relative intensities.

Sample preparation: If the polymers to be analyzed are in the molten form at room-temperature, spectra can be measured in transmission mode with the polymer spread out on a KBr tablet. A very thin film of the polymer is formed by first placing a small droplet of the polymer melt on top of the tablet, and then scraping off the excess polymer with a razor blade. Now the sample can be used for IR analysis. If the polymers solid at room-temperature, spectra can be measured in transmission mode with the polymer in the form of a 10-50 micrometer thick films. For this an appropriate amount of polymer is first dissolved in a suitable solvent. The solution is poured into a Petri dish. After evaporation of the solvent, the polymer films are carefully dried under vacuum at 50 °C. Now the sample can be used for IR analysis.

An IR spectrum is measured in the frequency range 400-4000 cm^{-1} in transmission mode. The result obtained as absorption spectra.

The following parameters can be evaluated from the IR spectra of the polymer

(i) Polymer Structure

Identify the peaks or bands in the spectra using tables of the absorption frequencies in wave numbers of various chemical entities as given below to assign the structure

Bond	Moiety	Type of vibration	Frequency in cm^{-1}	Intensity
C-H	Alkanes	Stretching	3000-2850	s
	CH_3	Bending	1450 and 1375	m
	CH_2	Bending	1465	m
	Alkenes	Stretching	3100-3000	m
		Bending out of plane	1000-650	s
	Aromatic	Stretching	3150-3050	s
		Bending out of plane	900-690	s
	Alkyane	Stretching	3300	s
	Aldehyde	Stretching	2900-2800 and	w
		Bending	2800-2700	w
C-C	Alkanes	Stretching	800-1300	w
C≡C	Alkenes	Stretching	1680-1600	m-w

	Aromatic	Stretching	1600-1475	m-w
C≡C	Alkyane	Stretching	2250-2100	m-w
C=O	Aldehyde	Stretching	1740-1720	s
	Ketone	Stretching	1725-1705	s
	Carboxylic acid	Stretching	1725-1700	s
	Ester	Stretching	1750-1730	s
	Amide	Stretching	1680-1630	s
	Anhydride	Stretching	1810-1760	s
	Acid chloride	Stretching	1800	s
C-O	Alcohol. Ether, Ester, Carboxylic acid, Anhydride		1300-1000	s
O-H	Alcohols, Phenols			
	Free		3650-3600	m
	H-bonded		3400-3200	m
	Carboxylic acids		3400-2400	m
N-H	Amines, Amides			
		Stretching	3500-3100	m
		Bending	1640-1550	m-s
C-N	Amines		1350-1000	m-s
C=N	Imines and oximes		1690-1640	w-s
C≡N	Nitriles		2260-2240	m
N=O	Nitro		1550 and 1350	s
S-H	Mercaptans		2550	w
S=O	Sulfoxides		1050	s
	Sulfones, Sulfonil chlorides, Sulfates, Sulfonamides		1375-1300 and 1350-1140	s
C-X	Fluoride		1400-1000	s
	Chloride		785-540	s
	Bromide,Iodide		< 667	s
-N=N-	Azo group		1570-1630	s

For illustration the interpretation of IR spectrum of polyvinyl alcohol is given below in Fig. 3.9.

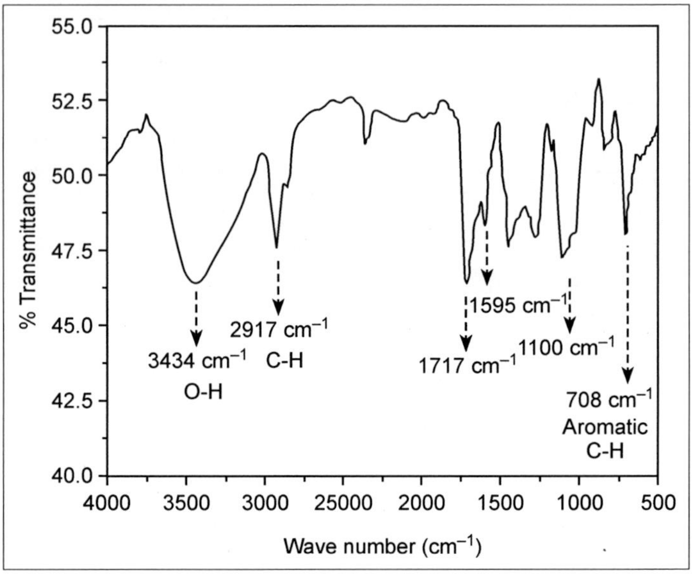

FIG. 3.9. IR spectrum of polyvinyl alcohol.

TABLE 3.0. Identified group frequencies in IR spectrum of polyvinyl alcohol.

Wave number (cm^{-1})	Assignment (PVA)
3434	O-H Stretching
2917	C-H asymmetric stretching (-CH_2)
1717	C=O Symmetric stretching *(Carbonyl group peak is due to absorption of the residual acetate used in the manufacture of PVA from hydrolysis of polyvinyl acetate)*
1595	C=C Stretching (aromatic)
1100	C-O Stretching
708	C-H Stretching (aromatic)
1432	C-H Symmetric bending (-CH_2)

(ii) Polymerization Kinetics

Infrared spectra of monomers differ markedly from spectra of the polymers. As a consequence, it is possible to use infrared spectroscopy to follow the course of polymerization reactions and to simultaneously analyze the structure of the polymer .Any chemical reaction, whether involving the main chain or side groups, results in a change of composition of one or more groups and consequently in the IR spectrum. This makes it possible to study oxidation, thermal degradation, cyclization, grafting, and other reactions of polymers. Evaluation of both qualitative and quantitative changes, as well as determination of kinetic constants of the reaction, is possible.

(iii) Crystallinity

In crystallization of polymers, the polymer forms crystalline and amorphous regions. The formation of crystalline regions is accompanied by an increase in new vibrational modes caused by their crystal lattice interactions. The IR spectrum of a given polymer differs by various absorption bands, depending on whether it is in the amorphous or crystalline state. The IR spectrum of crystalline polymer exhibits regularity in bands, splitting and frequency shifts. Other absorption bands are not affected by crystallization and remain the same in both cases. Crystalline and amorphous bands can be used in the determination of the degree of crystallinity.

3.3.2 NMR Characterization

Nuclear magnetic resonance (NMR) is a spectroscopic technique which involves the radio frequency electromagnetic radiations. The NMR technique utilizes the property of spin of the nuclei having odd number of protons and neutrons. Most common such nuclei are ^{1}H ^{13}C, ^{15}N, ^{17}O, ^{19}F etc. Among these ^{1}H NMR and ^{13}C NMR are most commonly used for structural determination. There is basic difference between NMR spectra of amorphous and highly crystalline systems. Highly organized and ordered molecular systems produce the spectra with very narrow signals while disordered materials are reflected by signals which are more or less broad depending on the degree of organization and assembling of the molecules. NMR is a well established, simple, fast, and fairly accurate method for the characterization of homo polymers and copolymers. The first step in applying this technique to polymers is to understand the spectral features, particularly the origin of each resonance observed. It has been applied for the determination of monomer sequence, chain configuration and polymer microstructure including stereo regularity and end-group composition. The interpretation of NMR spectra of polymer solutions generally follows the same approach as for smaller molecules. The number of signals in NMR spectrum signifies the number of equivalent sets of protons in a molecule. The positions of signals in the spectrum help us to know the nature of protons i.e. aliphatic, aromatic, acetylenic and vinylic adjacent to some electron attracting or electron withdrawing group. These protons have different electronic environment and thus they absorb at different applied strength, thus electronic environment of the proton decides the position of the absorption in the spectrum. The position of the absorption signal of protons is recorded relative to the signal of reference tetramethyl silane (TMS) protons at $\delta = 0$ or $\tau = 10$ ppm. Greater the deshielding of protons higher will be the value of δ and protons are said to be absorb downfield. Greater the shielding of protons smaller will be the value of δ and protons are said to be absorb upfield.

Because of the large size of a polymer chain relative to that of an organic molecule, the sample preparation procedure and spectrometer parameters used in acquiring the NMR spectra of polymers are slightly different from those used for small molecules. Because of the high viscosity of most polymer solutions, sample homogeneity can be a problem. Care must be taken to minimize concentration gradients and to ensure that no air bubbles are present. The simplest way to accomplish this is to let the samples stand for several hours, with or without heating. Another potential problem (for colored samples) arises from pigments that may contain paramagnetic impurities like iron oxide. During spectral accumulation, these magnetic particles migrate, causing fluctuations in the magnetic field, which leads to an extremely broad and becomes difficult to interpret the spectrum.

^1H NMR has also been applied in the determination of the molecular weight of the polymer. The area or intensity under the resonance peaks of the proton signal of a given species in the spectra is proportional to the amount of that species present in a given sample. This implies that the area of the ith peak is proportional to Ni, the number of molecules of the species "i" with molecular weight Mi, that is

$$Ai \propto NiMi$$

$$Ai = K\,NiMi$$

Where K is the constant of proportionality

Ai is the area or intensity of ^1H NMR

Ni is the number of molecules of species i

Mi is the molecular weight of species i

Now the equations for Number average and weight average molecular weight of polymer are given by,

$$M_n = \frac{N_1M_1 + N_2M_2 + N_3M_3 + N_4M_4 + \ldots + N_nM_n}{N_1 + N_2 + N_3 + N_4 + \ldots \ldots + N_n}$$

$$= \frac{\Sigma N_iM_i}{\Sigma N_i}$$

and

$$M_w = \frac{N_1M_1^2 + N_2M_2^2 + N_3M_3^2 + \ldots + N_n M_n^2}{N_1M_1 + N_2M_2 + \ldots + N_nM_n}$$

$$= \frac{\Sigma N_iM_i^2}{\Sigma N_iM_i}$$

Assuming K = 1, we get Ai = NiMi, Substituting this value in the equation of M_n and M_w, we get

$$M_n = \frac{\Sigma A_i}{\Sigma N_i} \quad \text{and} \quad M_w = \frac{\Sigma A_iM_i}{\Sigma A_i}$$

These equations establishes the relationship of the number-average molecular weight and weight average molecular weight with the ^1H NMR resonance peaks area or intensity of the detectable hydrogen atoms of a polymer.

In order to use this method, the following criteria must be fulfilled,

- Identifiable end-group protons distinguishable from repeating monomer-group protons by NMR.

- Accurate integration of one of the end-groups and repeating unit protons.
- Knowledge of formula weights of repeating monomer unit and both end groups.

Thus the method can be applied to determine the number-average molecular weight (Mn) value for polymer as follows.

The degree of polymerization (DP) or number of repeating units of a polymer is determined from its ^{1}H NMR spectrum by comparing the relative proton peak intensity of a known moiety (typically an end-group with a known number of protons) to that of the repeating chain unit of interest. We will derive the equation to calculate the DP as follows:

We have

$$A_i = K N_i M_i$$

$$K = \frac{A_i}{N_i M_i}$$

Where A_i is the area or intensity of the ^{1}H NMR peak of species i

N_i is the number of repeating units of species i

M_i is the number of protons of species i and

K is the constant.

Let us consider the ^{1}HNMR signals of two moieties, x (Repeating unit moiety) and y (End group moiety with known number of protons). Apply the above equation to both moieties x and y. We can write,

$$K_x = \frac{A_x}{N_x M_x} \quad \text{and} \quad K_y = \frac{A_y}{N_y M_y}$$

where A_x is the area or intensity of the ^{1}HNMR peak of repeating moiety x.

N_x is the number of repeating units of moiety x (*i.e.* DP).

M_x is the number of protons in each repeating unit of moiety x.

A_y is the area or intensity of the ^{1}H NMR peak of end group moiety y.

N_y is the number of end group moiety y and

M_y is the number of protons of end group moiety y.

For a given polymer, $K_x = K_y$, therefore we can write,

$$\frac{A_x}{N_x\, M_x} = \frac{A_y}{N_y\, M_y}$$

$$N_x = \frac{A_x\, N_y\, M_y}{M_x\, A_y}$$

Consequently, Mn can be calculated by substituting the value of N_x i.e. DP in the following equation,

$$M_n = (N_x * FW_x) + FW_y$$

where N_x is the number of repeating units or DP

FW_x is the formula molecular weight of each repeating unit and

FW_y is the total formula molecular weight of both end-groups.

Following example illustrates this method:

From the ^{1}HNMR spectra of the polymer methoxy poly (ethylene glycol) ethyl amine following data is generated.

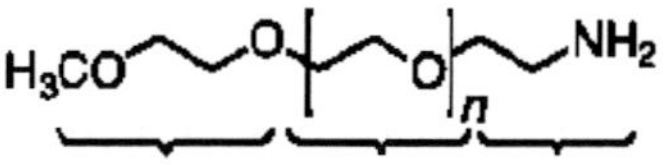

Formula Mol. wt. 75.1 44.06n 44.09 g mol^{-1}

FIG. 3.10. Structure of Methoxy poly (ethylene glycol) ethyl amine ($MPEGNH_2$) along with formula molecular weight of end groups (y) and repeating unit (x)

The protons of the methoxy (CH_3O-) and amino (-NH_2) end-groups resonated as a singlet and broad triplet respectively at 3.25 and 2.71 ppm (J = 5.6 Hz), whereas the prominent peak of the oxyethylene protons (-$(OCH_2CH_2)_n$-) was centered at 3.52 ppm. The peak area or intensity of the end-group CH_3O- is 0.9 and the peak area or intensity of repeating units - $(OCH_2CH_2)_n$ is 133.1

Determination of the Number of Repeating Units (DP):

To calculate the number of repeating units (LN_x) in the $MPEGNH_2$ chain, the peak areas of CH_3O- (-OMe) and -$(OCH_2CH_2)_n$- (-OEt-) were obtained from the ^{1}HNMR spectrum as given above are substituted into the equation,

$$N_x = \frac{A_x\, N_y\, M_y}{M_x\, A_y}$$

$$N_x = \frac{\text{Area of the peak of repeating moiety} \times \text{No. of end groups} \times \text{No. of protons of end group}}{\text{No. of protons of each repeating molety} \times \text{Area of the peak of end group moiety}}$$

$$N_x = \frac{133.1 \times 1 \times 3}{4 \times 0.9} = 110.92 \textit{ i.e. } \text{approx. } 111$$

Therefore the number average molecular weight of polymer will be equal to,

$$M_n = (N_x \times FW_x) + FW_y$$

$$M_n = (\text{DP * Formula mol. wt. of repreating unit}) + \text{Formula mol. wt. of both end groups}$$

$$M_n = (111 \times 44.06) + (75.1 + 44.09)$$

$$M_n = 5006.33 \text{ g per mol.}$$

3.3.3 X-Ray Diffraction Study (XRD)

Polymers may exist in different forms. They can be either partially crystalline or non-crystalline depending on how they are made and processed. Partially crystalline polymers are usually called crystalline polymers. The degree of crystallinity is an important parameter for crystalline polymers. The physical and mechanical properties of polymers are profoundly depends on the degree of crystallinity. The degree of crystallinity can be determined by various methods like XRD, NMR, IR and density measurements. Out of these X-ray diffraction is found to be the most superior and sophisticated non-destructive analytical method to evaluate the degree of crystallinity of polymers.

The crystalline non-polymeric materials comprises of repeating unit cells containing each of the component atoms in the material. Each unit cell is composed of one or more molecular units. In a polymer this is not possible, the molecules are the chains containing large number of repeating units from which it is made called monomer. The amorphous polymers are generally found in random coil conformation and have a disordered chain structure. This is the most common structure of many polymers. Crystalline polymers are predominantly in the all-trans conformation with their chains are arranged in lamellae, as shown below.

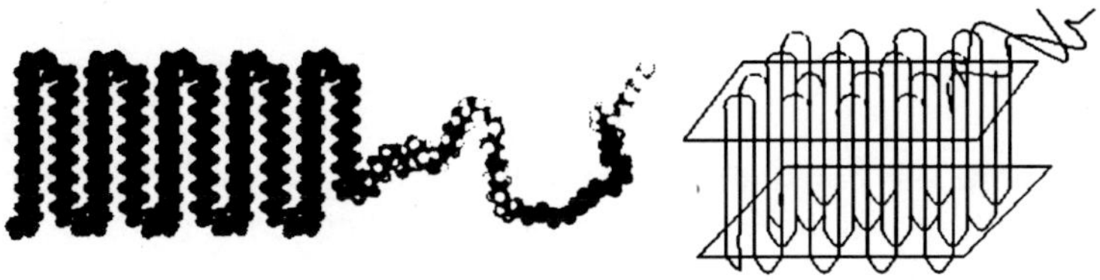

FIG. 3.11. lamellae chain structure of crystalline polymers

The polymer crystal made up from one dimensional chain folded sequence is shown in left where as two dimensional lamellar arrangement of folded chains is shown in the right of above figure 3.11.

As a result of difference between the amorphous and crystalline arrangements of polymer chains, the X-ray diffraction patterns of the two phases are very different. The amorphous phase contains no long order, meaning that there are no regular crystalline planes to diffract X-rays. Thus the incident X-rays are scattered randomly and there are no sharp peaks in the diffraction pattern. In the crystalline phases, the repeating lamellar chains provide a regular structure, thus the diffraction pattern will contain sharp, prominent signature peaks, the position of which depends on the exact spacing between chains. XRD uses total X-ray scattering of both crystalline and amorphous phases to determined the degree of crystallinity and no external standard is needed. The crystalline percentage in a polymer can be determine by exposure of high energy X-ray light on the sample to get the diffraction pattern. The diffraction pattern is divided in two parts. Sharp peaks are related to diffraction of crystallites where as broad are related to scattering of amorphous phases. The assumption can be made that the area under the peaks are proportional to the scattering intensities of crystalline and amorphous phases. The X-ray diffraction pattern of the polymer can be analyzed to get the following information.

(*i*) Percentage crystallinity: The percentage crystallinity (x_c) in XRD is determined by dividing the total area of crystalline peaks by the total area under all diffraction curves that is both crystalline and amorphous peaks. It can also be determined in terms of intensity of both crystalline and amorphous peaks.

$$x_c = \frac{I_{cr}}{I_{cr} + K I_{am}} = \frac{A_{cr}}{A_{cr} + K A_{am}}$$

Where K is constant related to the different scattering factors of crystalline and amorphous phases. For relative measure K =1.

$$x_c = \frac{\text{Total area of crystalline peaks}}{\text{Total area of all the peaks}}$$

Air scatter must be removed prior to profile fitting of peaks and air scatter is typically determined by running a 'blank' (i.e. collecting a data pattern under identical conditions without a sample) and subtracting that blank pattern from the pattern of sample under investigation. It should also be noted that the assignment and exact location of the amorphous and crystalline peaks are typically left to the judgement of the analyst.

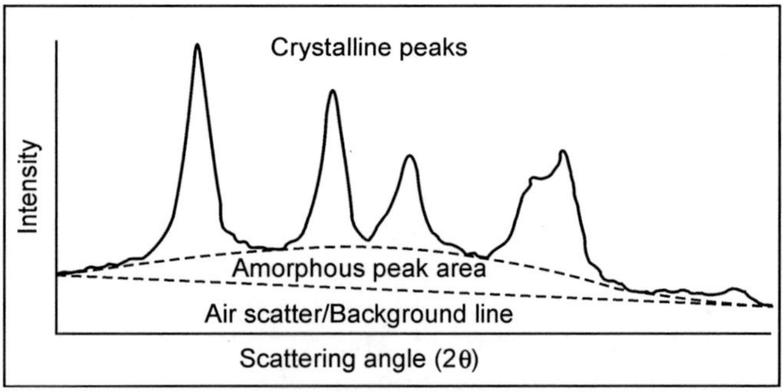

FIG. 3.12. Typical XRD pattern of polymer

(*ii*) **Crystallites dimensions:** The width of the diffraction peak at a height half way between background and the peak maximum (known as Full width at half maximum i.e. FWHM) is related to crystallites dimensions. The narrow half width corresponds to the bigger crystallites and the broad half width corresponds to smaller crystallites.

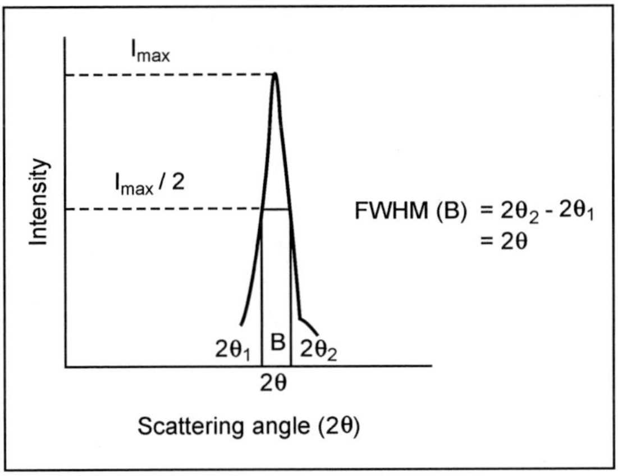

FIG. 3.13. Showing FWHM of crystalline peak

***(iii)* Microstructure:** Crystallite size in polymers can be determine by using Scherrer's equation given by,

$$L_{hkl} = \frac{K\lambda}{\beta \cos \theta}$$

Where L_{hkl} = Crystallite dimensions (in A°) along the direction perpendicular to the crystallographic plane *hkl*.

K = Constant dependent on crystallite shape (usually K = 0.89)

θ = Diffraction angle of the *hkl* reflection.

λ = Wavelength of X-ray used ($\lambda_{Cuk\alpha}$=1.5418 A°)

β = Full width at half maximum (FWHM) of peak related to the crystallographic plane *hkl* and can be determine as below,

The observed peak width (B) involves the peak broadening due to sample and instrumental effects too. Therefore actual peak width (β) can be determined by subtracting the instrumental broadening (b) from the observed peak width,

$$\beta = B - b$$

Where B = Full width at half maximum or integral breadth of peaks ($B = \Delta 2\theta = 2\theta_2 - 2\theta_1$)

b = Instrumental broadening taken as a full width at half maximum of a peak of low molecular weight crystalline compound.

❑❑❑

SECTION— II

4. Experimental Synthesis of Polymers

4

Experimental Synthesis of Polymers

This chapter describes the experimental procedure for the laboratory synthesis of common polymers. The various techniques used for the synthesis of polymers are already explained in chapter 2 of section-I.

EXPERIMENT NO. 1

AIM

Preparation of polystyrene by Bulk polymerization.

APPARATUS

250 ml three necked flask, Funnel, Magnetic stirrer, Reflux condenser, Nitrogen gas cylinder, Thermometer etc.

CHEMICALS

Styrene, Benzoyl peroxide (initiator), Dodecyl mercaptan (DDM), Methanol.

REACTION

Polystyrene is prepared by the bulk polymerization of styrene using benzoyl peroxide as an initiator at 70°C temperature. It involves the free radical polymerization mechanism. In bulk polymerization technique liquid monomer containing dissolved initiator is to be heated to get the polymer.

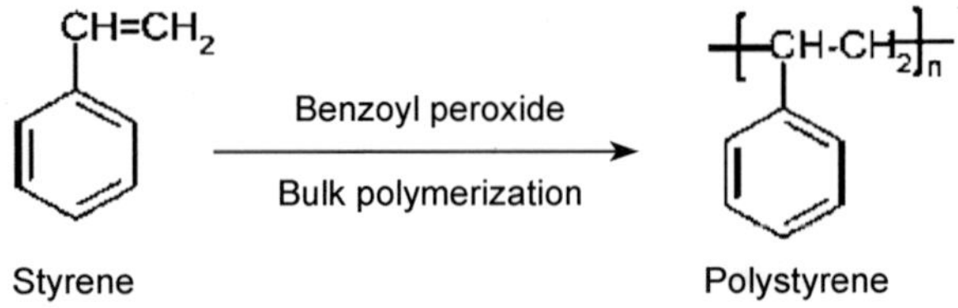

PROCEDURE

Take 50 mg of initiator benzoyl peroxide in a dry 250 ml three necked round bottom flask and dissolve it in 10 ml of freshly distilled styrene monomer. Now add 10 ml of styrene monomer containing 500 mg of dodecyl mercaptan to the flask. Flush the contents of the flask with oxygen free dry Nitrogen gas for about 10 min. Heat the flask in constant temperature water bath maintained at 70 °C using water condenser for about three hours or till to get the highly viscous solution. The beginning of polymerization reaction is indicated by the increase in the viscosity of the solution. When the viscosity of the solution reaches to the maximum, remove the flask from water bath. Allow it to cool to room temperature. Dissolve the solidified polystyrene by adding and stirring with 30 ml of toluene in the

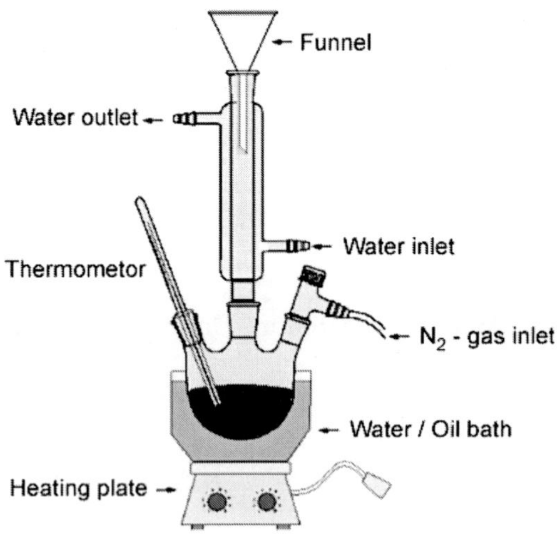

FIG. 4.0. Typical set up of three necked flask

flask. When polystyrene gets completely dissolve in toluene, transfer the solution slowly into 200 ml of non solvent methanol in a beaker with constant stirring. The polystyrene will precipitate out as a white fluffy solid. Collect the polymer by the process of filtration, wash it two times with methanol and dry it in an oven at 50 °C. Take the weight of dried polymer.

RESULT

Report the yield of polystyrene.

PROPERTIES

- It generally consists of linear molecules.
- It is transparent, hard and brittle in nature.
- It is chemically inert however acids, alkalis, oxidising or reducing agents have little effect on it.
- It has poor weathering properties and brittleness, therefore outdoor exposure causes yellowing and crazing of polystyrene.
- It has a low heat distortion temperature (85°C), therefore articles made of it cannot be sterilised with steam.

APPLICATIONS

- It is used in the manufacturing of moulded containers, lids, jars, bottles, decorative articles, radio and television cabinets, toys etc.

❑❑❑

EXPERIMENT NO. 2

AIM

Preparation of polystyrene by suspension polymerization.

APPARATUS

Three necked round bottom flask, Funnel, Magnetic stirrer, Reflux condenser, Nitrogen gas cylinder, Thermometer etc.

CHEMICALS

Styrene, Benzoyl peroxide (initiator), polyvinyl alcohol (PVA as a dispersing agent), Water (as a dispersion medium).

REACTION

Polystyrene can also be prepared by the suspension polymerization of styrene by using benzoyl peroxide as an initiator at 70°C temperature. It involves the free radical polymerization mechanism. In suspension polymerization technique liquid monomer containing dissolved initiator is suspended in aqueous dispersion medium in the form of droplets and heated to get the polymer.

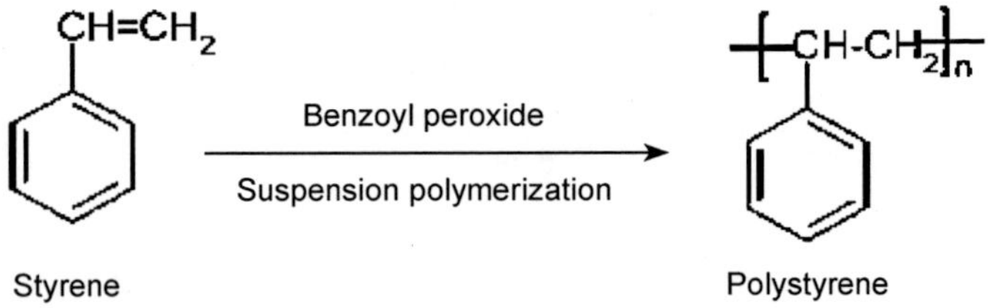

PROCEDURE

Take 50 ml of 2% aqueous polyvinyl alcohol solution in a three necked round bottom flask. Prepare the solution containing 0.5 g of bonzoyl peroxide in 10 ml of freshly distilled styrene. This solution of styrene and benzoyl peroxide is then transfer slowly and drop wise to the hot 50 ml 2% aqueous polyvinyl alcohol solution in three necked round bottom flask with continuous stirring for the period of about 15 min. Flush the contents of the flask with oxygen free dry Nitrogen gas for about 10 min. Heat the reaction mixture in a water bath maintained at 70 °C using water condenser with continuous stirring for about five hours. The polystyrene will appear in the form of beads in the reaction mixture. Filter off the polymer beads, wash with 2% HCl solution followed by water. Dry the polymer beads in an oven at 50 °C. Take the weight of dried polymer beads.

RESULT

Report the yield of polystyrene

APPLICATIONS

Refer experiment No. 1

PROPERTIES

Refer experiment No. 1

❑❑❑

EXPERIMENT NO. 3

AIM

Preparation of polystyrene by emulsion polymerization.

APPARATUS

Three necked round bottom flask, Funnel, Magnetic stirrer, Reflux condenser, Thermometer, Nitrogen gas cylinder, etc.

CHEMICALS

Styrene monomer, Potassium persulphate (initiator), Sodium lauryl sulphate (emulsifier), Sodium chloride (coagulant), Methanol, Water.

REACTION

Polystyrene can also be prepared by the emulsion polymerization of styrene by using potassium persulphate as an initiator at 70°C temperature. In emulsion polymerization technique liquid monomer forms a uniform emulsion in aqueous dispersion medium.

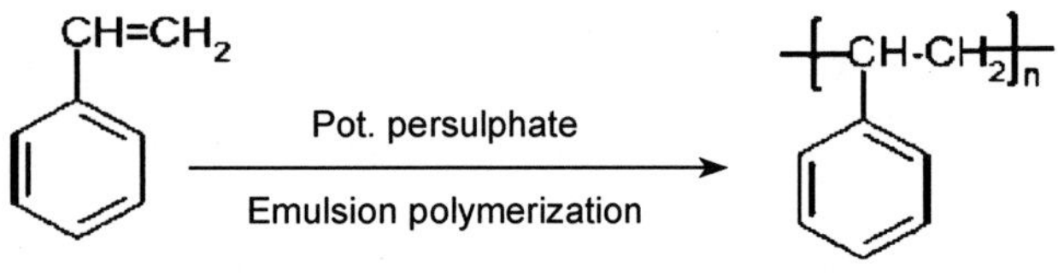

Styrene — Polystyrene

PROCEDURE

Take 100 ml. of water in a three necked round bottom flask. To this add 0.05 g. sodium lauryl sulphate and 0.05 g. potassium persulphate with stirring. Then add 50 ml. of freshly distilled styrene through a dropping funnel, slowly and drop wise over a period of about an hour. Flush the contents of the flask with oxygen free dry Nitrogen gas for about 10 min. Heat the contents of the flask on a water bath maintained at 70 °C with continuous stirring for about five hours. At the end, the content of the flask get converted into the latex. Stop heating the flask, Cool it, add 50 ml of 3% sodium chloride solution slowly and stir gently. The emulsion is broken and the latex coagulates to give the polymer polystyrene mass which is then filter off, wash with warm water followed by methanol. Dry the polymer in an oven at 60 °C. Take the weight of dried polymer latex.

RESULT

Report the yield of polystyrene.

APPLICATIONS

Refer experiment No. 1

PROPERTIES

Refer experiment No. 1

❑❑❑

EXPERIMENT NO. 4

AIM

Preparation of polymethyl methacrylate (PMMA) by suspension polymerization.

APPARATUS

Three necked round bottom flask, Funnel, Magnetic stirrer, Reflux condenser, Thermometer, Nitrogen gas cylinder, etc.

CHEMICALS

Methyl methacrylate, Benzoyl peroxide (initiator), Polyvinyl alcohol (dispersing agent), Distilled water (dispersion medium).

REACTION

Polymethyl methacrylate (PMMA) is prepared by the polymerization of methyl methacrylate by using benzoyl peroxide as an initiator at 80°C temperature. It involves the free radical polymerization mechanism. In suspension polymerization technique liquid monomer containing dissolved initiator is suspended in aqueous dispersion medium in the form of droplets and heated to get the polymer.

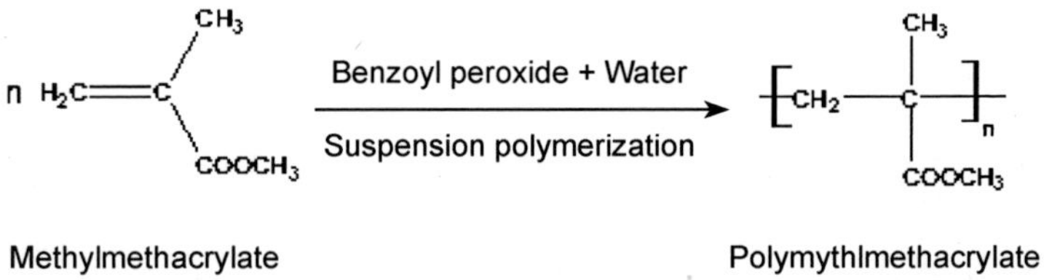

PROCEDURE

Take 10g of Methyl methacrylate (or equivalent volume) containing dissolved 0.1 g of benzoyl peroxide in a 250ml three necked round bottom flask equipped with magnetic stirrer, thermometer and water condenser. Add 20 ml of 2% aqueous polyvinyl alcohol solution which acts as a dispersing agent. Stir the solution vigorously. Flush the contents of the flask with oxygen free dry Nitrogen gas for about 10 min. Then heat the content of the flask at about 80°C temperature on a heating mental or water bath for 40 minutes. The internal temperature begins to rise and will reach a maximum of 85°C. At this point stop the heating and the solution is allowed to cool to 60°C and then add cold water to the flask. The granuals of polymer will get separate out, wash with water and dry in an oven at 60°C. Take the weight of dried granules.

RESULT

Report the yield of polymethylmethacrylate granules.

PROPERTIES

- It is colourless, transparent with excellent outdoor properties.
- It has good mechanical strength and chemical resistance.
- It is amorphous in nature due to presence of bulky side groups.
- It has good optical clarity.

APPLICATIONS

- It is used in the manufacturing of attractive signboards.
- It is used in buildings for decorative purpose.
- It can be used as an excellent substitute for glass in many applications.
- It is used in the manufacturing of contact lenses.

❑❑❑

EXPERIMENT NO. 5

AIM

Preparation of polyacrylamide by free radical polymerization.

APPARATUS

Three necked round bottom flask, Funnel, Magnetic stirrer, Reflux condenser, Thermometer, Nitrogen gas cylinder, etc.

CHEMICALS

Acrylamide monomer, Potassium persulphate (initiator), Methanol, Water

REACTION

Polyacrylamide is prepared by the free radical polymerization of acrylamide in aqueous medium by using potassium persulphate as an initiator at 70°C temperature. It is a water soluble polymer.

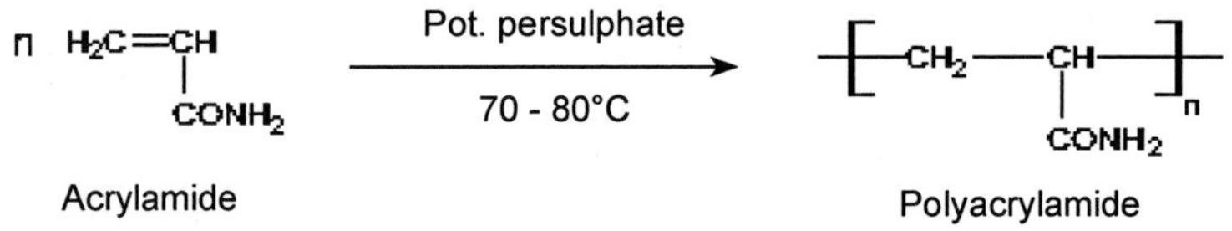

PROCEDURE

Dissolve 5 g of acrylamide in 100 ml of water in a three necked round bottom flask. Add 6 ml of 1 % potassium persulphate solution from dropping funnel, slowly and drop wise over a period of 10 minutes. Flush the contents of the flask with oxygen free dry Nitrogen gas for about 10 min. Heat the contents of the flask gently on a heating mental or water bath maintain at 65 °C with continuous stirring for about one hour. Stop the heating and transfer the reaction mixture drop wise to a beaker containing methanol with stirring to precipitate out the polymer. Separate the polymer by the process of filtration then wash with methanol and dry in an oven at 50 °C. Take the weight of dried polymer.

RESULT

Report the yield of polyacrylamide.

PROPERTIES

- Polyacrylamides are high molecular weight water soluble or swellable, non-ionic and biocompatible polymers.
- The cross-linked polyacrylamides can absorb and retain extemely large amounts of water because the amide groups form strong hydrogen bonds with water molecules.
- Polyacrylamides increases the viscosity of water and encourages the flocculation of particles present in water.
- Polyacrylamides have the glass transition temperature well above the room temperature.

APPLICATIONS

- The hydrated polyacrylamide is a soft gel that is used in gel electrophoresis as a super water-absorbing polymer.
- Polyacrylamide and its derivatives are widely used as thickening agent, binder, super absorbent, soil conditioner, filtering aid, flocculating agent, crosslinker, suspending agent, lubricant and oil recovery agent.
- Polyacrylamides also used in waste water treatment. When added to waste water, it causes suspended particles to aggregate and to precipitate.
- Another common use is oil extraction and recovery from water.

❑❑❑

EXPERIMENT NO. 6

AIM

Preparation of polyacrylamide by redox polymerization.

APPARATUS

Three necked round bottom flask, Funnel, Magnetic stirrer, Reflux condenser, Nitrogen gas cylinder, Thermometer etc.

CHEMICALS

Acrylamide, Ferrous ammonium sulphate (FAS), Hydrogen peroxide, Methanol, Water etc.

REACTION

Polyacrylamide can also be prepared by the free radical polymerization of acrylamide monomer in aqueous solution using a redox initiator at room temperature. Ferrous ammonium sulphate (FAS) in combination with hydrogen peroxide (Fenton's reagent) can used as the redox initiator.

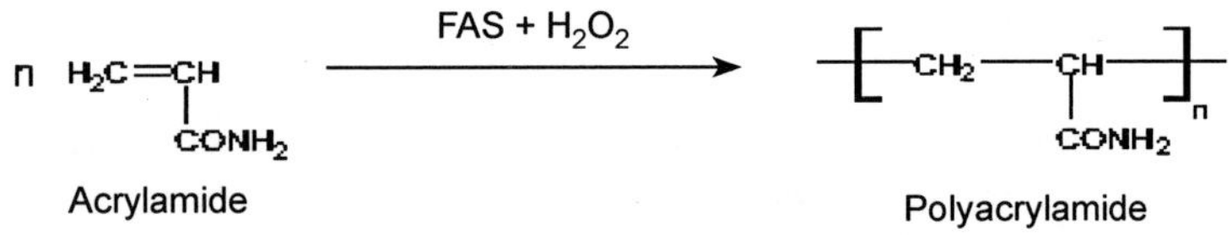

PROCEDURE

Take 10 g of pure acrylamide and dissolve it in 100 ml of distilled water in a 250 ml three necked round bottom flask. Flush the contents of the flask with oxygen free dry Nitrogen gas for about 10 min. To this add 25 ml of 0.5 M aqueous solution of ferrous ammonium sulphate (FAS) and 25 ml of 0.1 M aqueous solution of hydrogen peroxide with constant stirring for about 30 to 40 minutes. Add the resulting viscous solution with constant stirring to 250 ml of methanol containing 5 to 10 drops of HCl. The formation of light brown colored precipitate occurs. Filter the precipitate and dissolve it in 50 ml of water. Make the solution ammonical for the separation iron from the polymer in the form of ferric hydroxide. Filter off the ferric hydroxide precipitate. Then add the filtrate *i.e.* polymer solution drop wise with constant stirring to 100 ml of methanol. The polymer will get separate out as white precipitate. Allow it to settle for 10 minutes. Filter, dry in an air and take the weight of dried polymer.

RESULT

Report the yield of polyacrylamide.

APPLICATIONS

Refer experiment No. 5

PROPERTIES

Refer experiment No. 5

❑❑❑

EXPERIMENT NO. 7

AIM

Preparation of polyacrylonitrile (Polyvinyl cyanide) by suspension polymerization.

APPARATUS

Three necked flask, Funnel, Magnetic stirrer, Reflux condenser, Thermometer etc.

CHEMICALS

Acrylonitrile, Potassium persulphate (initiator), Sodium bi-sulphite.

REACTION

Polyacrylonitrile is prepared by the free radical polymerization of acrylonitrile by the free radical polymerization of acrylonitrile using potassium persulphate and sodium bi-sulphite as an initiator at 40°C.

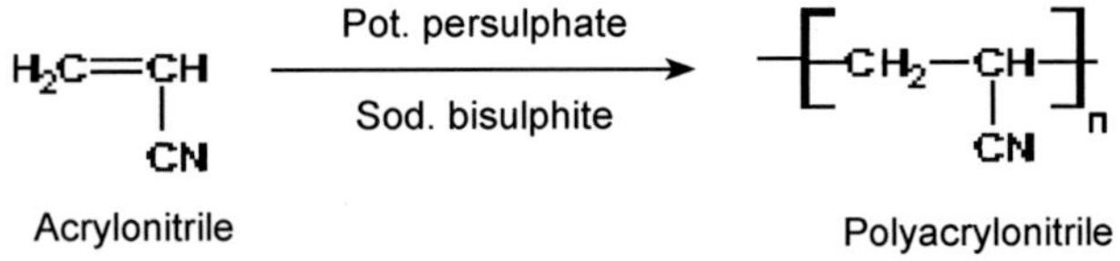

PROCEDURE

In 250ml three necked flask equipped with magnetic stirrer, thermometer and condenser on heating mental take 75ml of freshly boiled distilled water. Add 5.5g (or equivalent volume) of acrylonitrile slowly with stirring for 10 minutes. Reflux the solution gently to attain the temperature of 40°C. To this solution then add 75mg of potassium persulphate and then after one minute add 40mg of sodium bi-sulphite (These acts as an initiators of polymerization reaction). Heat the mixture with continuous stirring for three hours at 35°C to 40°C. The colorless aqueous solution becomes somewhat opalescent and white polymer begins to precipitate within 5 minutes. After heating filter the solid product, wash with water and dry it at 60°C in a vacuum oven. Take the weight of dried polymer.

RESULT

Report the yield of polyacrylonitrile.

PROPERTIES

- It has excellent resistance to heat up to about 220°C.
- It has very good mechanical properties.

APPLICATIONS

- It is used to produce fibres having wide applications.
- The copolymer of acrylonitrile with butadiene *i.e.* nitrile rubber is a material of great industrial importance.

❑❑❑

EXPERIMENT NO. 8

AIM

Preparation of polyacrylonitrile (Polyvinyl cyanide) by precipitation polymerization.

APPARATUS

250 ml Three necked flask, Separating funnel, Magnetic stirrer, Reflux condenser, Constant temperature bath, Thermometer, Nitrogen gas cylinder, Oven etc.

CHEMICALS

Acrylonitrile, Benzoyl peroxide (initiator), Toluene.

REACTION

Polyacrylonitrile can also be prepared by the free radical polymerization of acrylonitrile using benzoyl peroxide as an initiator at 70°C.

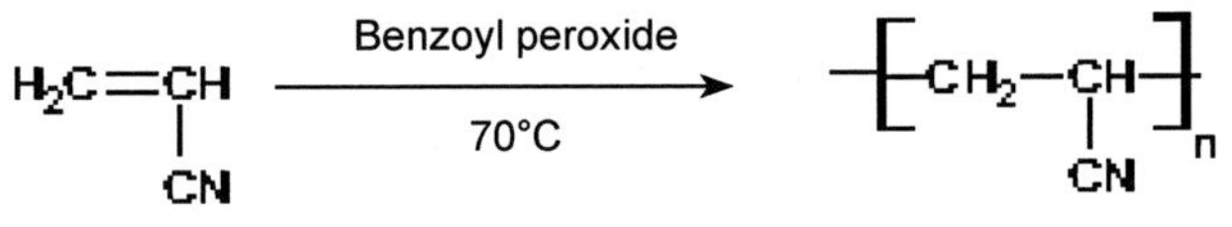

Acrylonitrile   Polyacrylonitrile

PROCEDURE

Take 50 mg of Benzoyl peroxide in 250 ml three necked round bottom flask .Dissolve it in a 10 ml of acrylonitrile monomer and then add 100 ml of toluene to the flask. Flush the contents of the flask with oxygen free dry Nitrogen gas for about 10 min. Place the flask in a constant temperature bath maintained at 70°C .The beginning of polymerization reaction is indicated by the appearance of turbidity and the formation of white precipitate of polymer at the bottom of the flask. This is due to the fact that the polymer formed is insoluble in its monomer as well as in toluene. Filter the precipitate of polyacrylonitrile, wash two three times with water and dry in an oven at 60 °C. Take the weight of dried polymer.

RESULT

Report the yield of polyacrylonitrile.

APPLICATIONS

Refer experiment No. 7

PROPERTIES

Refer experiment No. 7

❑❑❑

EXPERIMENT NO. 9

AIM

Preparation of polyethylacrylate by emulsion polymerization.

APPARATUS

Three necked round bottom flask, Funnel, Magnetic stirrer, Reflux condenser, Nitrogen gas cylinder, Thermometer etc.

CHEMICALS

Ethyl acrylate, Potassium persulphate (Initiator), Sodium lauryl sulphate (emulsifier), Methanol (coagulant), Water.

REACTION

Polyethylacrylate can be prepared by the polymerization of ethylacrylate by using potassium persulphate as an initiator.

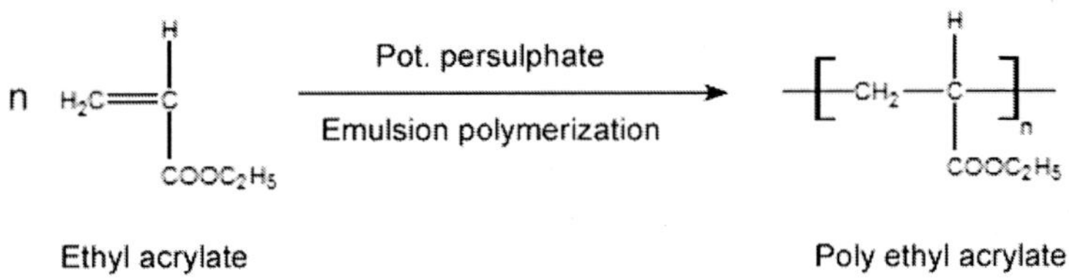

PROCEDURE

Ethyl acrylate (Boiling Point 99.5°C) is commercially available with 0.02 to 0.1 weight percent of polymerization inhibitor hydroquinone. It is essential to reduce the level of inhibitor hydroquinone to use ethylacrylate as a monomer in the synthesis polyethylacrylate. This can be achieved by vacuum distillation of ethyl acrylate. If the vacuum distillation is not possible then the level of inhibitor hydroquinone can be reduced by extraction with alkali. For this, wash two to three times 100 ml of ethyl acrylate with fresh 50 ml of 5% sodium hydroxide solution in a separating funnel with vigorous shaking. The monomer is then again washed repeatedly with water to make it free from alkali. The monomer is then allowed to stand over the granular calcium chloride till it gets free from turbidity and moisture. It is then filtered and stored in cold condition till further used. ***Do not pour ethyl acrylate directly into the sink.***

To prepare the Polyethyl acrylate, first form the monomer emulsion by dissolving 2 g of Sodium lauryl sulphate in 100 ml of water followed by slowly addition of 10 ml of monomer ethylacrylate with continuous stirring in a 250 ml three necked round bottom flask fitted with stirrer, water condenser and thermometer. Flush the contents of the flask with oxygen free dry Nitrogen gas for about 10 min. To this add 5 ml of 1% potassium persulphate solution with slowly and continuous stirring. Reflux the contents of the flask on a water bath at 75°C for about half an hour. Allow the content to cool. Treat the content of the flask with Methanol to coagulate the product. Filter the product and then wash with water. Dry and take the weight of polymer.

RESULT

Report the yield of polyethylacrylate.

PROPERTIES

- It is tough, rubbery polymer materials with little or no tack at room temperature.
- It is known for their high transparency and elasticity and have fairly good heat resistance up to 450 Kcal under dry heat.
- It also has good weather ability and ozone resistance since they do not have double bonds in the backbone.
- Polyethyl acrylate is often copolymerized with polymethyl acrylate to achieve the desired hardness, flexibility and strength.

APPLICATIONS

- Major applications are coatings, paints, textiles, leather finishing, automotive products, tape adhesives, and oil-resistant and high-temperature-resistant elastomers.

- It is also used as co-monomer to increase the plasticity of rigid and brittle plastics.
- It is used as pressure sensitive adhesives.
- Acrylic elastomers are often a cheap alternative to fluorocarbon polymers, silicones and fluoro silicones for elevated temperature applications (< 430 K).
- It is extensively used in the automotive industry for hoses, seals, gaskets and dampers that must function under long-term exposure to elevated temperatures and hydrocarbon oils.

❑❑❑

EXPERIMENT NO. 10

AIM

Preparation of polyglycerol phthalate (Glyptal resin).

APPARATUS

Round bottom flask, Funnel, Thermometer, etc.

CHEMICALS

Phthalic anhydride, Glycerol, Round bottom flask, Reflux condenser, Oil bath.

REACTION

Glycerol when heated with phthalic anhydride at 150-160°C, it undergoes esterification to produce polyglycerol phthalate. It is also known as glyptal resin.

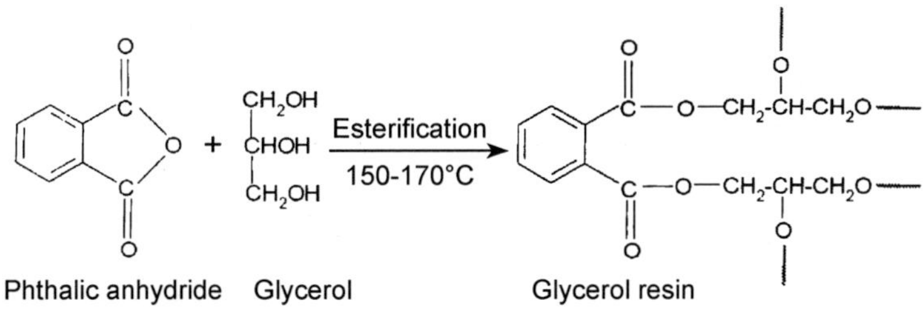

PROCEDURE

Take 7.4 g of phthalic anhydride and 3.07 ml of glycerol in a 250 ml Round bottom flask. Mix well the contents of the flask with the help of glass red. Attach an air condenser to the neck of flask. Reflux the content of the flask on oil bath or sand bath at 150 °C to 170 °C for about two hours. Transfer the brown yellow resin into a previously weighed test tube. Note down the weight of glyptal resin formed.

RESULT

Report the yield of glyptal resin.

PROPERTIES

- It has good surface wetting properties.
- It has good weathering properties

APPLICATIONS

- It is used in making protective furniture and architectural coatings with good weathering properties
- It is important ingredients in many synthetic paints, varnishes and enamels due to its versatility and low cost.
- It also finds applications as thermosetting plastics that can be molded.
- The principal applications are coatings, product finishing, special-purpose coatings and automotive refinishing primers.

❑❑❑

EXPERIMENT NO. 11

AIM

Preparation of phenol formaldehyde resin (Bakelite).

APPARATUS

Three necked round bottom flask, Funnel, Magnetic stirrer, Reflux condenser, Nitrogen gas cylinder, Thermometer etc.

CHEMICALS

Phenol, 37 % Formaline solution, Oxalic acid, Distilled water.

REACTION

If the reaction is carried out in acidic condition (oxalic acid) with stoichiometrically insufficient quantity of the formaldehyde only linear phenol formaldehyde molecules are formed. The resin formed can be stored for any period of time without getting harden and can be further cross linked when ever required by adding and heating with excess of formaldehyde. This type of phenol formaldehyde resins is known as Novlak or Bakelite resins.

If the reaction is carried out in alkaline condition (Sodium hydroxide) with stoichiometrically equi-molar quantities of phenol and formaldehyde with the controlled reaction rate and extent of conversion, it leads to the formation of linear phenol formaldehyde molecules with methylol groups at the ortho and para positions and therefore on standing or storage, slow reaction will continue leading to the hardening of the resin. This type of resin is called resol.

PROCEDURE

Take 32.5 g phenol and 23 ml of 37 % formaline and 1.0 g Oxalic acid dissolved in 10 ml of water in a 250 ml three necked round

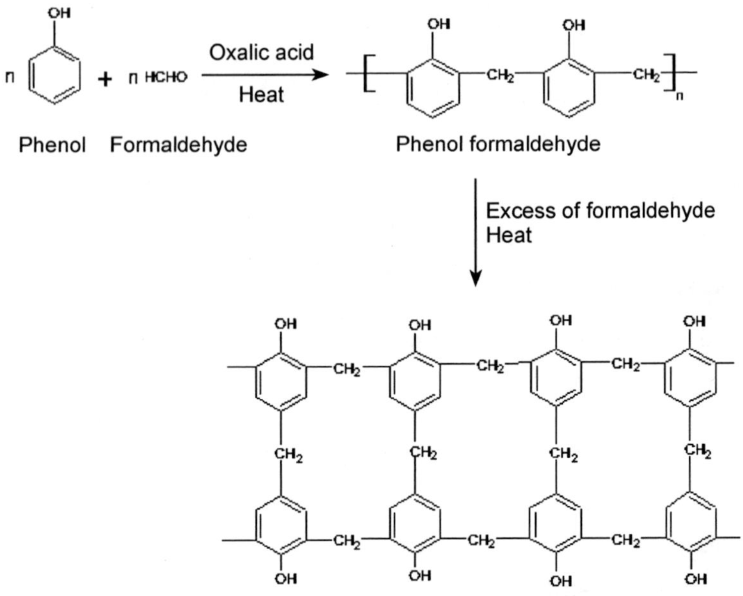

Cross linked phenol formaldehyde

bottom flask. Reflux the content of the flask to 60-80°C for about one and half hour using water condenser on a water bath. As the bubbles start to rise, showing the exothermic reaction has begun, the burner can be taken away and the liquor is allowed to boil on its own. Continue heating the flask to the desired time. A time will reach when the bubble will escape very difficultly from the surface of the solution due to the high viscosity of the solution. At that time stop heating and pour immediately the viscous content of flask in previously weighed test tube. Allow it to cool and note down the weight of resin formed.

RESULT

Report the yield of Phenol formaldehyde resin.

PROPERTIES

- It is thermosetting type of polymer.
- It has high heat and acid resistance.
- It has good electrical properties.

APPLICATIONS

- Phenol formaldehyde resins compounded with fillers like asbestos powder or sawdust are used for moulding electrical items, telephone instruments etc.
- The resins are used in the production of cook ware handles.
- The resins are used for making laminates and also as coating compositions also.

❑❑❑

EXPERIMENT NO. 12

AIM

Preparation of urea formaldehyde resin.

APPARATUS

Three necked round bottom flask, Funnel, Magnetic stirrer, Reflux condenser, Nitrogen gas cylinder, Thermometer etc.

CHEMICALS

Urea, 37 % formaline solution, 10 % NaOH aqueous solution

REACTION

Formaldehyde reacts with urea in acidic or alkaline medium to form urea formaldehyde resin. The reaction involves the formation of mono and dimethylol derivatives of urea which further undergoes condensation reaction with urea or themselves to form the urea formaldehyde resin.

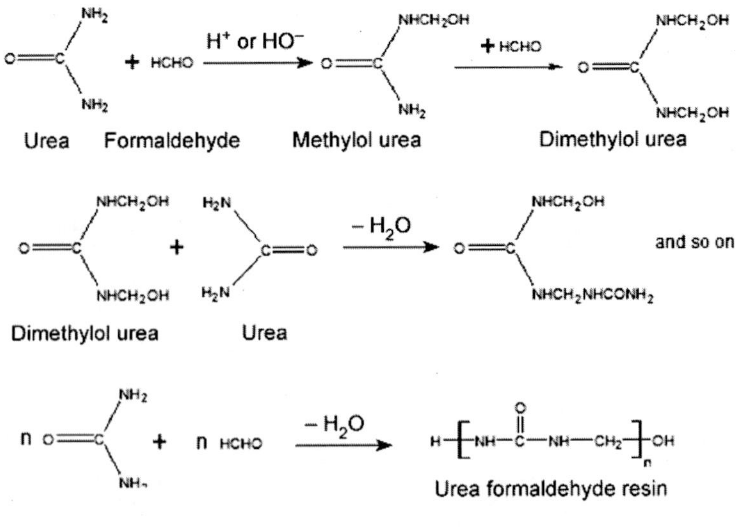

If the linear polymer is heated subsequently with excess of formaldehyde then the hydrogen atoms of the imide group in the linear polymer can be further replace with methylol groups and a three dimensional network of the polymer will be produced with the following structure.

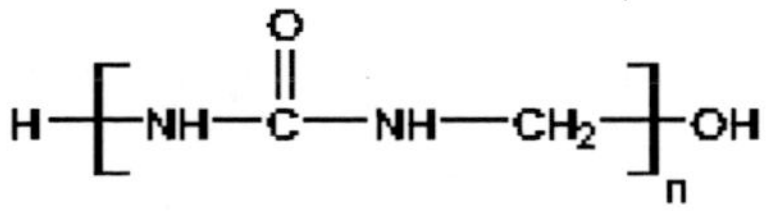

Urea formaldehyde resin

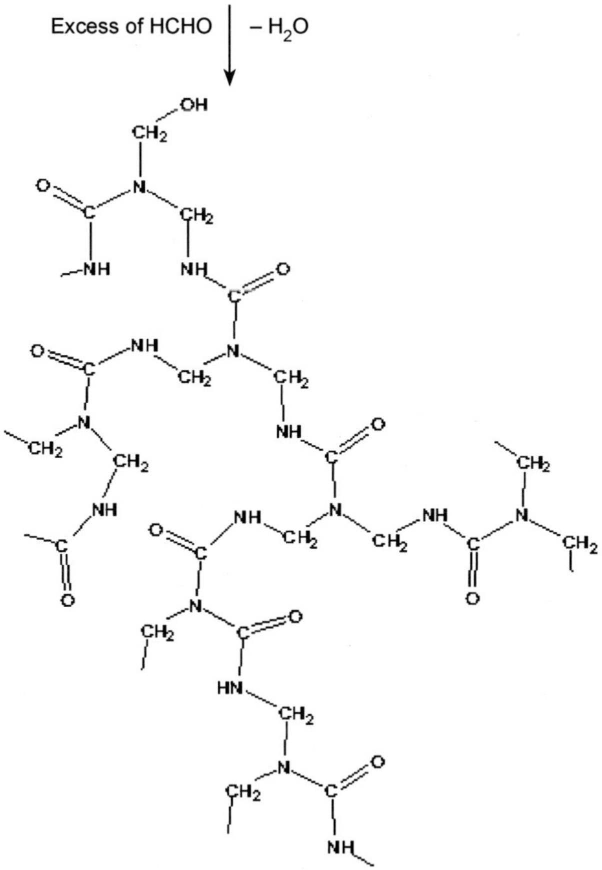

Cross linked urea formaldehyde resin

PROCEDURE

Urea formaldehyde resin can be prepared in acidic or alkaline medium as given below.

ACIDIC CONDITION

Take 10 g of urea and 13 ml of 37 % formaline solution in a 250 ml beaker. Stir the solution using magnetic stir to get homogeneous solution. Add 3, 4 drops of dilute hydrochloric acid with constant stirring and reflux for one to two hours. Then add 50 ml of water. The urea formaldehyde resin formed as white powder. Note down the weight of resin formed.

ALKALINE CONDITION

Take 13 ml of 37 % formaline solution in a 250ml three necked round bottom flask fitted with stirrer and reflux condenser. To this add drop wise 10 % Sodium hydroxide aqueous solution to get the pH of about 7.5, which can be confirmed by using pH paper. Then add 10 g of urea to it and reflux the content of the flask on water bath for about one hour. It is an exothermic reaction hence care should be taken that the content of the flask are agitated while refluxing the solution. After refluxing, allow it to cool to room temperature to get pre polymer. Acidify the product formed with 0.5 ml of acetic acid and heat further for five minutes at 100°C to remove the water content and to get the urea formaldehyde resin. Transfer the resin in a previously weighted test tube. Note down the weight of resin formed.

RESULT

Report the yield of Urea formaldehyde resin.

PROPERTIES

- Uncross-linked urea formaldehyde resins are water soluble.
- These resins have excellent electrical properties.
- As compare to phenolic resins, these are light in colour and acid proof.

APPLICATIONS

- Since uncross-linked urea formaldehyde resins are water soluble these are useful as sizing agents and textile-finishing resins.
- This resin is used in paper industry to improve the wet strength of paper.
- This resin is used in plywood industries as an adhesive.
- This resin is used to prepare electrical switches.

❑❑❑

EXPERIMENT NO. 13

AIM

Preparation of melamine formaldehyde resin.

APPARATUS

Three necked round bottom flask, Funnel, Magnetic stirrer, Reflux condenser, Thermometer etc.

CHEMICALS

Melamine, 37% formaline, 10% NaOH, Glycerine, Ethyl alcohol.

REACTION

Melamine reacts with formaldehyde in basic medium to form trimethylol and hexamethylol derivatives of melamine. The

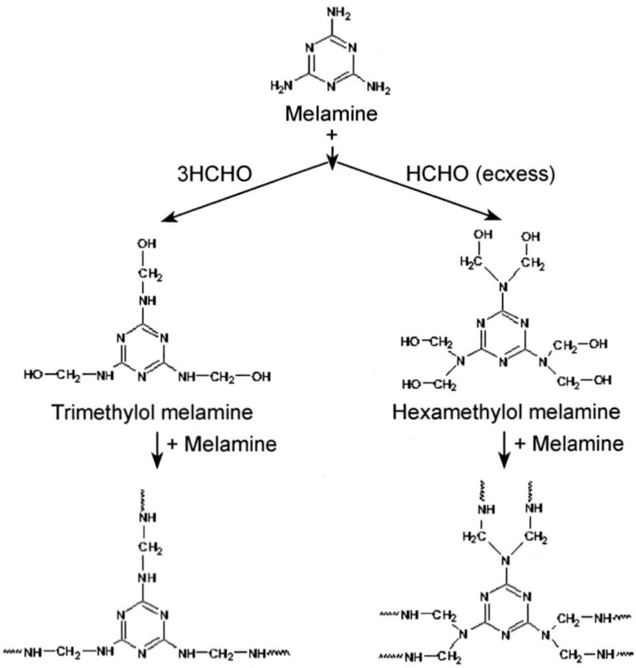

Cross linked urea formaldehyde resin

polycondensation of these methylol derivatives with melamine will results in the formation three dimensional cross linked structure of melamine formaldehyde resin, which is found to be insoluble in most of the solvents.

PROCEDURE

Take a clean and dried 250 ml three necked round bottom flask equipped with a stirrer, reflux condenser and a thermometer. Add about 15 g of melamine and 35 g of 37% formaldehyde solution into the flask and stir the reaction mixture well. Then add a few drops of 10% sodium hydroxide to adjust the P^H of reaction mixture to 8.5. This can be checked by using P^H paper. Heat the content of the flask on water bath at 80 °C for about 30 minutes till to get the homogeneous solution. During the course of heating check the P^H periodically and maintain it at 8.5. Add about 5 g of glycerin (Softening agent) to it and continue the stirring of reaction mixture for another 15 minutes. A viscous liquid is obtained, which on further heating converts into clear hard melamine formaldehyde resin. Wash it with ethyl alcohol and note down the weight of resin formed.

RESULT

Report the yield of Melamine formaldehyde resin.

PROPERTIES

- Uncross-linked melamine formaldehyde resins are water soluble.
- These resins have excellent electrical properties.
- As compare to phenolic resins, it is light colored and more clearer.
- Tensile strength and hardness of this resin is much better than the phenolic resins.
- Heat and moisture resistance are less as compare to phenolic resins but better than urea formaldehyde resins.

APPLICATIONS

- This resin finds use in manufacture of decorative laminates.
- It is used for the preparation of lacquers.
- This resin is used to prepare electrical goods.

❑❑❑

EXPERIMENT NO. 14

AIM

Preparation of Aniline formaldehyde resin.

APPARATUS

Three necked round bottom flask, Beaker, Funnel, Magnetic stirrer, Reflux condenser, Thermometer etc.

CHEMICALS

Aniline, 37 % formalin solution, Dil. Hydrochloric acid, 1% Sodium hydroxide solution, Distilled water.

REACTION

Aniline and formaldehyde reacts to form linear aniline formaldehyde resin which further in presence of excess of formaldehyde convert into cross linked aniline formaldehyde resin through methylene bridges.

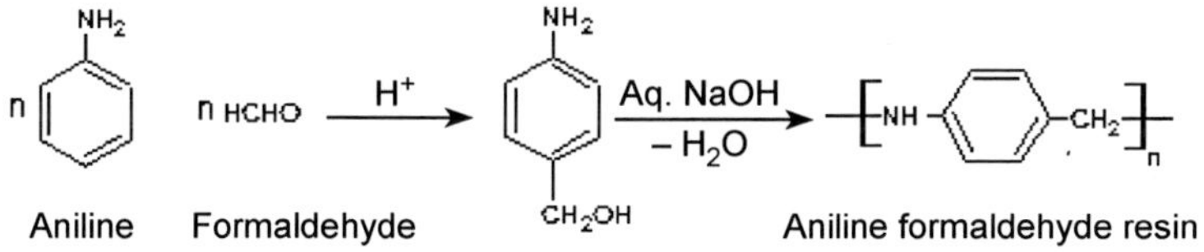

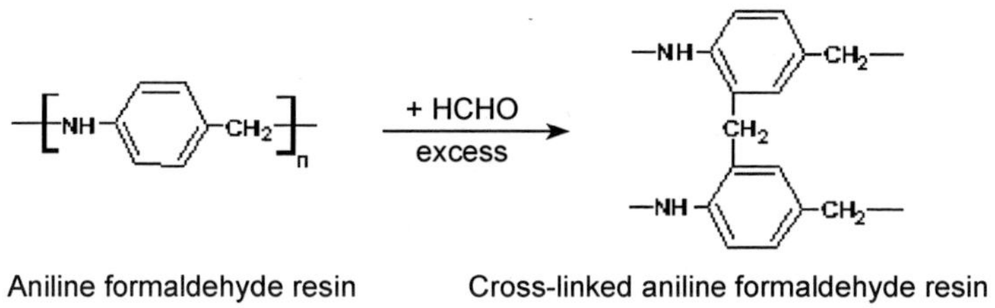

PROCEDURE

Take 9.3 g of aniline and 100 ml of dilute hydrochloric acid in a 250ml beaker. Add 10 ml of 37 % formaldehyde solution to it with constant stirring. Stir the mixture well for about 15-20 minutes to form an orange red solution. Transfer this solution in 100 ml of aqueous 1% sodium hydroxide solution placed in a three necked round bottom flask equipped with water condenser and stirrer. Reflux the content of the flask on water bath for about 45 minutes. Aniline formaldehyde resin will precipitate out. Filter and wash the product with water. Dry it and take the weight of dry resin.

RESULT

Report the yield of Aniline formaldehyde resin.

PROPERTIES

- It is yellow to red translucent high viscosity liquids or solids thermoplastic resin.
- It has well weather-resistant, UV-resistance, water-resistance, oil resistance and alkali resistance.

APPLICATIONS

- Aniline formaldehyde resins are used in molded products.
- It is also used as adhesives in laminated construction materials, such as plywood.

❑❑❑

EXPERIMENT NO. 15

AIM

Preparation of cellulose nitrate from cellulose using Nitrating mixture.

APPARATUS

Beaker, Conical flask, Funnel, Thermometer etc.

CHEMICALS

Mercerized cotton, Concentrated HNO_3, Concentrated H_2S0_4, Acetone, conical flask, beaker, Ice bath and thermometer.

REACTION

Cellulose is a naturally occurring linear, stereo-regular polysaccharide made up of β D (+) Glucose residues. Nitration of cellulose by a mixture of strong nitric acid, sulphuric acid and a small quantity of water gives cellulose nitrate polymer. It is also known as Nitro cellulose. At higher temperature the reaction rate increases but it leads to the chain degradation, therefore the nitration is carried out at about 20°C temperature.

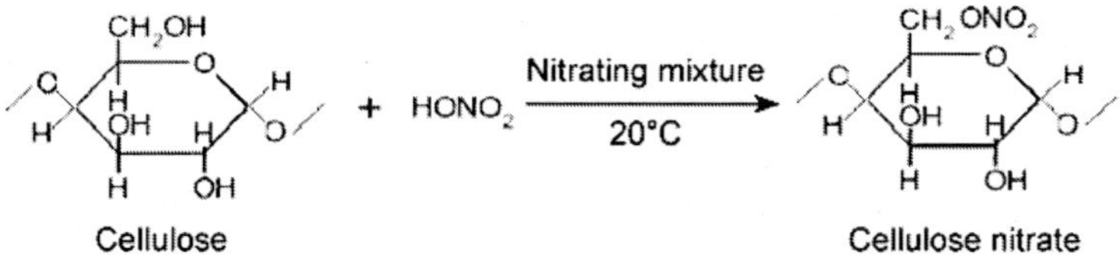

PROCEDURE

Prepare a nitrating mixture by taking 8 ml water and 60 ml of concentrated Nitric acid (HNO_3) in a clean dry conical flask. Cool the flask in ice bath to 10°C and add 60 ml of concentrated sulphuric acid (H_2SO_4) in such a way that temperature should not rise above 20°C, now add 1g of mercerized cotton piece into the nitrating mixture. Place the flask for 3 hours at room temperature with occassional stirring. After three hour, remove the cotton from nitrating mixture, wash it with water to make it free from acid. This can be confirmed by the test with litmus paper. Dissolve the cotton in minimum quantity of acetone in a beaker. Pour the solution of acetone into the ice cold water taken in another beaker. Precipitates of cellulose nitrate are separated out, filter and wash it with cold water followed by alcohol. Dry it and determine the weight of the polymer formed.

RESULT

Report the yield of Cellulose nitrate polymer.

PROPERTIES

- It has excellent impact resistance and is one of the most inexpensive plastics available.
- It has the disadvantage of having very high inflammability.

APPLICATIONS

The applications of cellulose nitrate depends on the degree of nitration as given below,

- Cellulose nitrate with a nitrogen content more than 12.3% is used as explosive (*e.g*. gun-cotton).
- Cellulose nitrate with a nitrogen content 10.5% to 11% is used as celluloid plastic.
- Cellulose nitrate with a nitrogen content 11% to 12% is used as notrolacquers.
- Cellulose nitrate with a nitrogen content 12% to 13.5% is used as photographic films.
- It is used to make moulded decorative articles, fountain pens, frames for spectacles etc.

❑❑❑

EXPERIMENT NO. 16

AIM

Preparation of cellulose triacetate from cellulose.

APPARATUS

Beaker, Conical flask, Funnel, Thermometer etc.

CHEMICALS

Mercerized cotton, Concentrated HNO_3, Concentrated H_2S0_4, Acetone, conical flask, beaker, ice bath and thermometer.

REACTION

Cellulose is a naturally occurring linear, stereo-regular polysaccharide made up of βD (+) Glucose residues. Another important cellulosic plastic is cellulose acetate. It is produced by the reaction of acetic anhydride with cellulose in presence of sulphuric acid or perchloric acid as a catalyst. The product obtained by complete acetylation of all hydroxyl groups in each gluciside ring of cellulose is called primary cellulose acetate or cellulose triacetate.

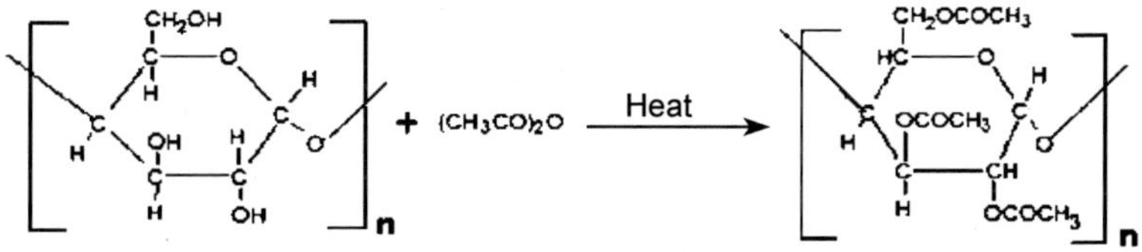

Cellulose Acetic anhydride Cellulose triacetate

PROCEDURE

Take 2 g of cotton in a beaker. Add 72 ml of glacial acetic acid to it. Mix cotton thoroughly with the glacial acetic acid. Place the beaker in a water bath for about 30 minutes with constant stirring at 50-55°C. Remove the beaker from water bath and allow it to cool to 30-35°C. Then add 48 ml of acetic anhydride and 2 ml of sulphuric acid. Reflux the mixture at temperature 50-

55°C on water bath for about one and half hour with constant stirring to get the clear solution. It should be noted that the temperature should not rise above 55°C to avoid the degradation of cellulose chains. After heating, pour the content of the flask in large quantity ice cold water. Cellulose triacetate get precipitate out. Filter and wash the product with water to remove acetic acid and finally dry it in an oven. Take the weight of the dry polymer.

RESULT

Report the yield of Cellulose triacetate polymer.

PROPERTIES

- It has good optical clarity.
- It has good moisture resistance.

APPLICATIONS

- It is used for the manufacture of non-flammable safety cinematographic films, moulded articles and fibres.
- It is used as photographic films.

❏❏❏

EXPERIMENT NO. 17

AIM

Preparation of Polysulphide rubber (Thiokol).

APPARATUS

Three necked round bottom flask, Funnel, Magnetic stirrer, Reflux condenser, Thermometer etc.

CHEMICALS

Powder sulpher, Sodium hydroxide, 1, 2-dichloroethane, Distilled water.

REACTION

Polysulphide rubber Thiokol is a synthetic rubber. It can be prepared by the reaction between 1, 2-dichloroethane and excess amount sodium tetrasulphide in presence of sodium hydroxide.

$$Na_2S + 3S \longrightarrow Na_2S_4$$

Sodium sulphide   Sodium powder   Sodium tetrasulphide

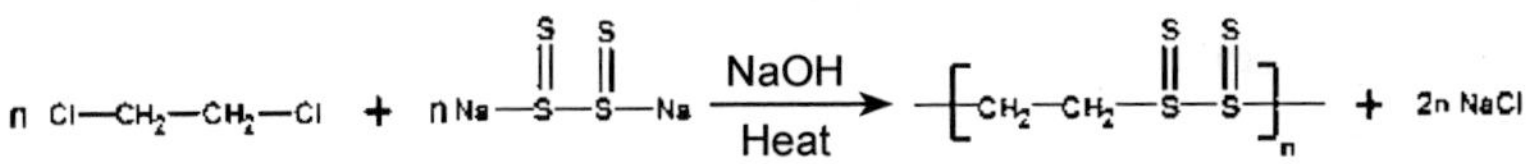

1,2 Dichloroethane   Sodium tetrasulphide   Poly sulphide rubber

PROCEDURE

Take 10 g of sodium hydroxide in a 250ml of round bottom flask. Dissolve it in 100 ml of water. Add 7 g sulphur powder to this in small amounts with constant stirring. Reflux the solution on water bath using water condenser for about one hour till the sulphur powder get dissolve completely. If some sulphur remains undissolved filter the solution. Cool the red brown solution to 60-70°C and add 8 ml of 1, 2-dichloroethane drop wise with constant stirring. The solution is further reflux at 80°C for about one hour or until yellow colored rubber polymer separates out as a lump in the solution. Cool the content to room temperature and collect the Thiokol rubber. Wash it with water under the tap. Dry it with the help of filter paper and take the weight of dry polymer.

RESULT

Report the yield of Polysulphide rubber

PROPERTIES

- The physical properties of the polysulphides vary with the length of the aliphatic group and the number of sulphur atoms present in the chain.
- Polysulphides have excellent oil resistance properties.
- These have very low gas permeability and are practically unaffected by ozone.
- These have very slow aging.
- These are excellent fuel materials.

APPLICATIONS

- These are mainly used to make sealants, gaskets, balloons, fabric coatings and gasoline hoses.
- These materials on mixing with inorganic oxidisers such as ammonium perchlorate, can be used as solid propellants for rockets.

❑❑❑

EXPERIMENT NO. 18

AIM

Preparation of Polyvinyl alcohol by the alcoholysis of polyvinyl acetate

APPARATUS

Three necked round bottom flask, Funnel, Magnetic stirrer, Reflux condenser, Thermometer etc.

CHEMICALS

Polyvinyl acetate, anhydrous methanol, anhydrous potassium hydroxide, Distilled water, P^H paper

REACTION

Polyvinyl alcohol can be prepared by the hydrolysis of polyvinyl acetate in presence of alkali

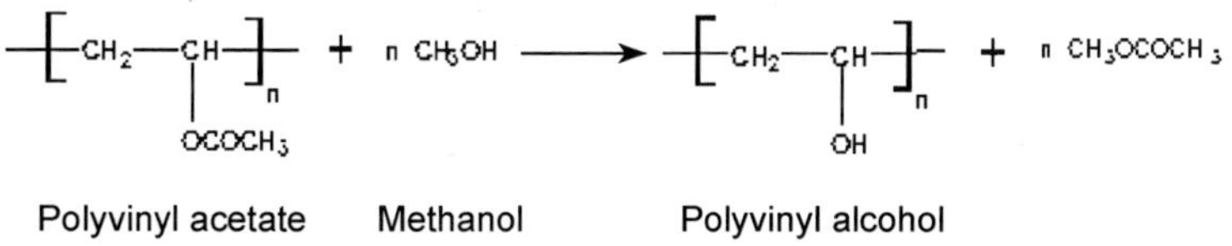

PROCEDURE

Take 1 g of anhydrous potassium hydroxide in the form of fine powder in a 250 ml conical flask. Add 100 ml of methanol in it and stopper the flask to preserve the solution till further used. Now take 10 g of polyvinyl acetate and 100 ml of anhydrous methanol in a 250 ml three-necked round bottom flask fitted with a mechanical stirrer, a reflux condenser and funnel. Reflux the content of the flask on water bath for about 20 minutes with vigorous stirring to dissolve the polyvinyl acetate completely. Reduce the flame of the burner and continue the refluxing at moderate rate. Then add 20 ml of potassium hydroxide solution in methanol prepared earlier, drop by drop through the funnel. It is important to note that, the temperature of the reaction system should not increase above 60°C during the addition of methanolic potassium hydroxide solution. After the complete addition, continue the refluxing for 2 hours. Polyvinyl alcohol separates out as a fine powder. Cool the flask. Filter off the polyvinyl alcohol formed. Wash the product repeatedly with anhydrous methanol until residual alkali has been removed from the product. The further purification can be achieved by dissolving the washed polyvinyl alcohol in 100 ml of distilled water by heating at 80°C. The solution is then poured drop wise in 200 ml of anhydrous acetone with vigorous stirring. The precipitate of polyvinyl alcohol is then filtered using suction pump and dried in a desiccators and take the weight of dry polymer.

RESULT

Report the yield of polyvinyl alcohol

PROPERTIES

- Unlike many polymers it is water soluble at about 90°C.
- It is good protective colloid for aqueous emulsions.
- Like starch solution, polyvinyl alcohol solution gives an intense blue colour with iodine.

APPLICATIONS:

- It is used as a protective colloid for aqueous emulsions and suspensions.
- The fibres of polyvinyl alcohol on treatment with a concentrated aqueous solution of sodium sulphate containing calculated amounts of formaldehyde and sulphuric acids to convert the hydroxyl groups in polymer chains into formal groups can be used as a substitute for cotton fibres.

❑❑❑

EXPERIMENT NO. 19

AIM

Preparation of Nylon 6 from caprolactum.

APPARATUS

Three necked round bottom flask, Beaker, Funnel, Magnetic stirrer, Reflux condenser, Nitrogen gas cylinder, Thermometer etc.

CHEMICALS

Adipic acid, Caprolactum, Methanol, 1% Sodium hydroxide solution, Distilled water.

REACTION

Nylon polymers are the aliphatic polyamides. There are different types of nylon polymers, which are usually differentiated by the number system, depending on the number of carbon atoms present in the monomer molecules. The nylon 6 is prepared by the ring opening polymerization of caprolactum at 240-270 °C in presence of water to form amino caproic acid which further undergo self poly condensation to form nylon 6.

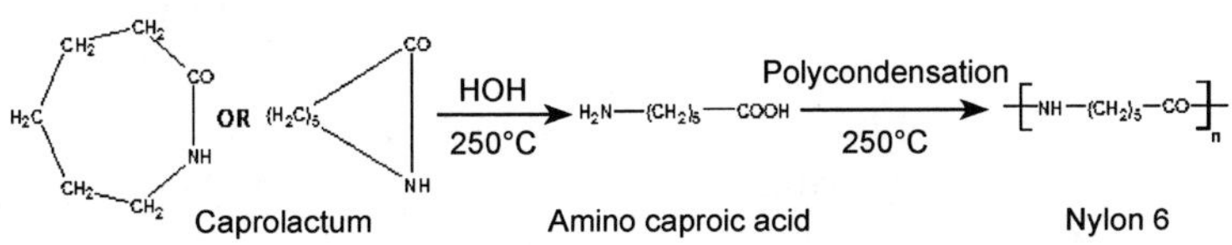

PROCEDURE

Take 8 g of pure caprolactum in a Pyrex polymerization reaction tube together with 1 ml of water. Seal the tube with rubber septum. Inject two syringe needles (one long and one short) into it. Pass a stream of oxygen free nitrogen gas slowly through the long needle for about 5 minutes to remove air through a short needle. Place the tube in an oil bath maintain at 250°C for about two hours. The melting of monomer caprolactum indicates the beginning of polymerization reaction. After the completion of heating, take the tube out of the oil bath and remove the needles from it. Immediately cool the tube under running water by occasionally shaking which helps in the formation of polymer layer on the sides of the tube. Hold the polymerization tube with thick clean cloth piece and break carefully by using hammer. Separate out the polymer from the glass surface. Crush the product into fine powder and take the weight of polymer. Store it in an air tight bottle.

RESULT

Report the yield of Nylon 6 resin

PROPERTIES

- It is a synthetic fiber and generally white in color but can be dyed to in a solution bath prior to production for different color results.
- It has good tensile strength, abrasion resistance and oil resistance.
- It is insoluble in many solvents.
- It is semicrystalline polyamide.

APPLICATIONS

- It is used in the manufacture of electrical and electronic components.
- Nylon 6 finds application in a broad range of products requiring materials of high strength.
- It is widely used for gears, fittings and bearings in automotive industry for under the hood parts and as a material for power tools housings.
- Nylon 6 is used as thread in bristles for toothbrushes, surgical sutures and strings for acoustic and classical musical instruments like guitars, sitars, violins, violas and cellos.
- It is also used in the manufacture of a large variety of threads, ropes, filaments, nets and tire cords as well as hosiery and knitted garments.

❑❑❑

EXPERIMENT NO. 20

AIM

Preparation of Nylon 6, 6 from adipic acid and hexamethylene diamine.

APPARATUS

Three necked round bottom flask, Beaker, Funnel, Magnetic stirrer, Reflux condenser, Nitrogen gas cylinder, Thermometer etc.

CHEMICALS

Adipic acid, Hexamethylene diamine, Ethanol, Distilled water.

REACTION

Nylon 6, 6 is the aliphatic polyamide. It is prepared by the poly condensation reaction between dicarboxylic acid (Adipic acid) and diamine (Hexamethylene diamine) with the elimination of water molecule. The high molecular weight polymer can be obtained by maintaining the stoichiometric equivalence of the monomers. The first number 6 in nylon 6, 6 polymer represent the number of carbon atoms in diamine monomer and the second number 6 represent the number of carbon atoms in dicarboxylic acid monomer.

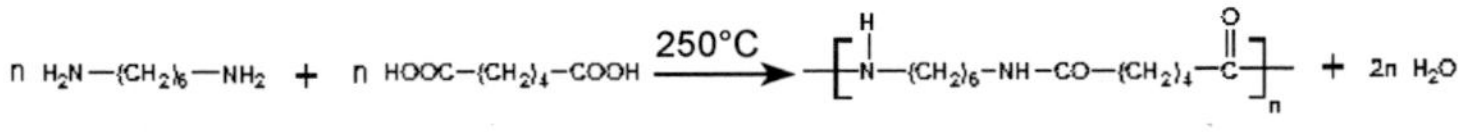

Hexamethylene diamine | Tetramethylene dicarboxylic acid (Adipic acid) | Nylon 6 6

PROCEDURE

Preparation of Hexamethylene Diammonium Adipate *i.e.* Nylon Salt

Dissolve 14.4 g of adipic acid in 120 ml of 95% ethanol in a beaker and filter it. In another beaker dissolve 11.6 g of Hexamethylene diamine in a mixture of 32 ml of 95% ethanol and 6 ml of distilled water and filter it. Now transfer the Hexamethylene diamine solution slowly and drop wise to the beaker containing adipic acid solution with vigorous stirring using magnetic stirrer for about 20 minutes. During the course of addition the rise in temperature of reaction system takes place due to the exothermic neutralization reaction. Continue the stirring for about 30 minutes even after the complete addition. The precipitation of white crystalline nylon salt will be occurred with fall in the temperature of reaction system. Allow the precipitate to settle down. Decant out the supernatant solution. Wash the precipitate in the same beaker by repetitive decantation with 95 % ethanol. Filter the solution under suction and air dry the precipitate.

Preparation of Nylon 6, 6

Take 10 g of above prepared nylon salt and 0.055 g of adipic acid in a Pyrex reaction tube. Evacuate the tube by connecting it to a vacuum pump for about 5-10 minutes. Flush the tube with oxygen free nitrogen gas and re evacuate it again. Seal the tube under vacuum carefully. Place the sealed tube in an oil bath maintain at 250°C for about 2-3 hours. After the heating, the tube is allowed to cool. Hold the tube with thick clean cloth piece and break carefully by using hammer. Separate out the translucent plug of Nylon 66 polymer from the glass surface. Take the weight of polymer. Store it in an air tight bottle.

RESULT

Report the yield of Nylon 6, 6 resin.

PROPERTIES

- It has good tensile strength, abrasion resistance and toughness up to 150°C.
- Insoluble in many solvents.
- It has good oil resistance.
- It is a plastic as well as fibre.

APPLICATIONS

- It is used to produce tyre cord.
- It is used to make monofilaments and ropes.
- It is used to make textile fibres for use in dresses, undergarments.
- It is used as a good substitute for metals in gears and bearings

❑❑❑

EXPERIMENT NO. 21

AIM

Preparation of Nylon 6, 6 by interfacial polycondensation method.

APPARATUS

Beaker, Funnel, Magnetic stirrer, Forceps.

CHEMICALS

Adiponyl chloride, Hexamethylene diamine, Cyclohexane, Sodium bicarbonate, Distilled water.

REACTION

Nylon 6, 6 can also prepared by the interfacial poly condensation reaction between Adiponyl chloride and Hexamethylene diamine with the elimination of Hydrochloric acid which dissolves in water.

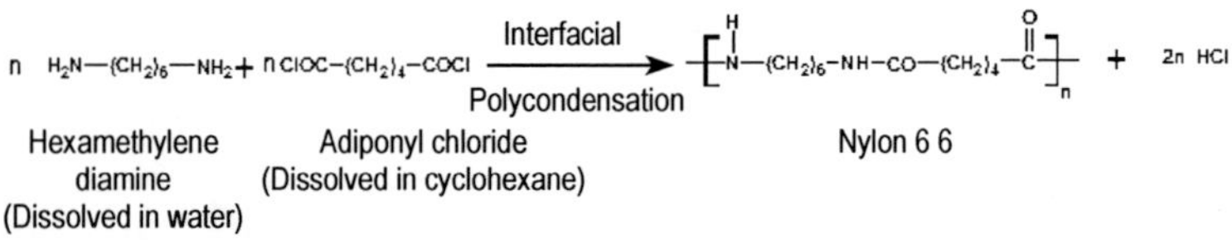

PROCEDURE

Dissolve 10 g of adiponyl chloride in 100 ml cyclohexane in one beaker and 10 g of hexamethylenediamine along with 2 g of sodium bicarbonate in 100 ml of water in another beaker. Add adiponyl chloride solution carefully and slowly to hexamethylenediamine solution. These two liquids do not mix, and two layers will form. The product nylon 6 6 will formed at the interface in the form of thin film while the byproduct hydrochloric acid dissolves in water and neutralized by sodium bicarbonate. Remove the product formed continuously by picking up the thin film using forcep.

RESULT

Report the yield of Nylon 6, 6 film.

APPLICATIONS

Refer experiment No. 21

PROPERTIES

Refer experiment No. 21

❑❑❑

EXPERIMENT NO. 22

AIM

Preparation of Nylon 6, 10 by interfacial polycondensation method.

APPARATUS

Beaker, Funnel, Magnetic stirrer, Forceps.

CHEMICALS

Sebacoyl chloride, Hexamethylene diamine, Dichloromethane, Sodium bicarbonate, Distilled water.

REACTION

Nylon 6, 10 can be prepared by the interfacial poly condensation reaction between sebacoyl chloride and hexamethylene diamine with the elimination of hydrochloric acid which dissolves in water. The polymer forms at the interface between two layer of aqueous phase and organic phase. The aqueous phase also contains sodium bicarbonate so as to neutralized the byproduct HCl. The denser organic phase forms the lower, while the aqueous phase will be carefully poured on top of the organic phase. There is no need to control of the stoichiometric ratio in this experiment. The stoichiometry at the interface is controlled perfectly by the mass transfer controlled rate of diffusion and consumption of each monomer species.

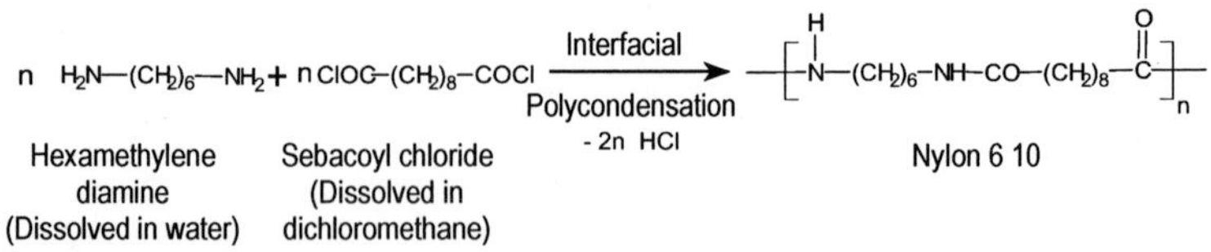

PROCEDURE

Dissolve 10 g of Sebacoyl chloride in 100 ml of Dichloro-methane in one beaker and 10 g of hexamethylenediamine along with 2 g of sodium bicarbonate in 100 ml of water in another beaker. Add hexamethylenediamine solution carefully and slowly to Sebacoyl chloride solution. These two liquids do not mix and two layers will form. The product nylon 6, 10 will formed at the interface in the form of thin film while the byproduct hydrochloric acid dissolves in water and get neutralized by sodium bicarbonate. Remove the product formed continuously by picking up the thin film using forcep for further formation of fresh film at the interface.

RESULT

Report the yield of Nylon 6, 10 film.

PROPERTIES

- Nylon-6,10 is semicrystalline polyamide.
- Nylon 6,10 is similar in most properties to nylon 6 and 6, 6.
- It has a lower melting point and better resilience than that of Nylon 6 and 6, 6. Nylon 6,10 is most often used for paintbrushes, toothbrushes and molded plastic items.

APPLICATIONS

- It is commonly used in monofilament form in applications such as bristles and brushes.
- Due to its low moisture absorption compared to other nylons, it retains its properties better when wet.

❑❑❑

EXPERIMENT NO. 23

AIM

Preparation of epoxy resin from Bisphenol – A and Epichlorohydrine. (Diglycidyl ether of bisphenol-A, DGEBA).

APPARATUS

Three necked round bottom flask, Beaker, Funnel, Magnetic stirrer, Reflux condenser, dropping funnel, Vacuum pump, Thermometer etc.

CHEMICALS

Bisphenol – A, Epichlorohydrin, 10% and 40% sodium hydroxide, Benzene, Distilled water.

REACTION

Epoxy resins are basically polyethers. The most common epoxy resin is prepared by the reaction of bisphenol – A with

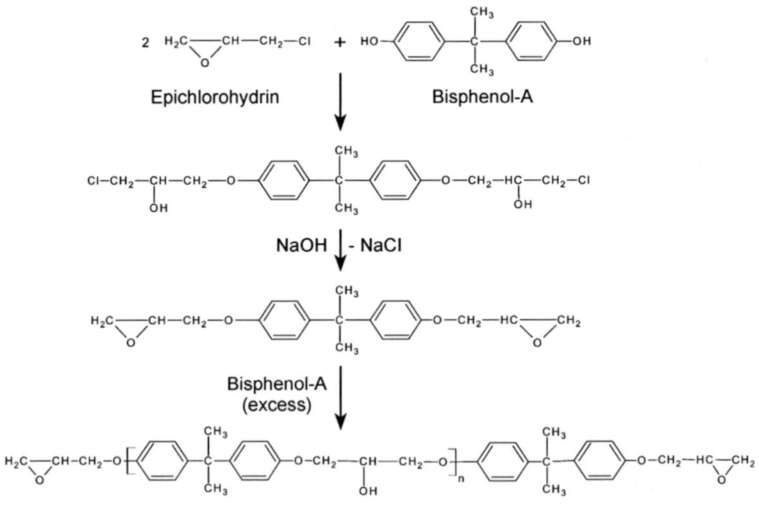

Epoxy resin (DGEBA)

excess of epichloro hydrine in alkaline condition. Bisphenol A or bis (4-hydroxy phenyl) dimethylmethane (BPA) is a colourless solid (mol.wt 228), where as Epichlorohydrin (ECH) is a colourless liquid with the characteristic odour of many chlorinated hydrocarbon solvents. Epoxy resin obtained through this reaction is either highly viscous liquid or solid with high melting point.

PROCEDURE

Liquid Epoxy resin

Take 10 g of bisphenol – A and 6 g of epichlorohydrin in a three necked round bottom flask equipped with water condenser, thermometer and dropping funnel. Heat the content of the flask till the epichlorohydrin begins to boil. Stop the heating. Add 20 ml of 40 % aqueous sodium hydroxide solution slowly and drop wise for the period of half an hour. The reaction should allow to subside before adding more alkali to it. After the complete addition of alkali continue refluxing the mixture for about an hour. Stop the Heating when the reaction mixture turned viscous. Cool the flask to the room temperature using cold water bath. Reflux the content of the flask under vacuum at 70 °C so as to remove the non reacted epichlorohydrin and water from the reaction mixture. Stop refluxing and add 30 ml of benzene to precipitate out the byproduct sodium chloride salt. Remove the salt by vacuum filtration and wash with 20 ml of benzene. Remove the excess of benzene from the filtrate completely by vacuum distillation. The resulting clear pale yellow and highly viscous liquid is a epoxy resin.

Solid Epoxy resin

Take 10 g of bisphenol-A and 4 ml of 10 % aqueous sodium hydroxide solution in a three necked round bottom flask equipped with water condenser, thermometer and dropping funnel. Heat the content of the flask to 45°C for about 15 minutes. Then add 6.5 g of epichlorohydrin rapidly and reflux the mixture on water bath at 80°C for about one and half hour. The solid product separates out in the reaction mixture. Filter the product, wash it repeatedly with hot water to remove the alkali completely. Dry the resin in vacuum at 130°C for 48 hours. Take the weight of dry resin.

RESULT

Report the yield of epoxy resin

PROPERTIES

- These epoxy resins have remarkable chemical resistance and good adhesive properties.
- These have good dimensional stability and mechanical properties.
- When epoxy resins properly cured, yield very tough material.

APPLICATIONS

- These are excellent structural adhesives.
- The cured epoxy resins are used in industrial floorings, foams, potting materials for electrical insulations etc.
- These epoxy composites are used in aerospace and aircrafts.

❑❑❑

EXPERIMENT NO. 24

AIM

Preparation of Aniline epoxy resin from Aniline and Epichlorohydrin. (Diglycidyl aniline epoxy resin, DGA).

APPARATUS

Three necked round bottom flask, Beaker, Funnel, Magnetic stirrer, Reflux condenser, Dropping funnel, Vacuum pump, Nitrogen gas cylinder, Vacuum oven, Thermometer etc.

CHEMICALS

Aniline, Epichlorohydrin, 40% sodium hydroxide, Benzene, Methanol, Distilled water.

REACTION

Diglycidyl aniline (DGA) is prepared by the condensation of aniline with excess of epichlorohydrine in alkaline condition using azeotrophic mixture of solvents.

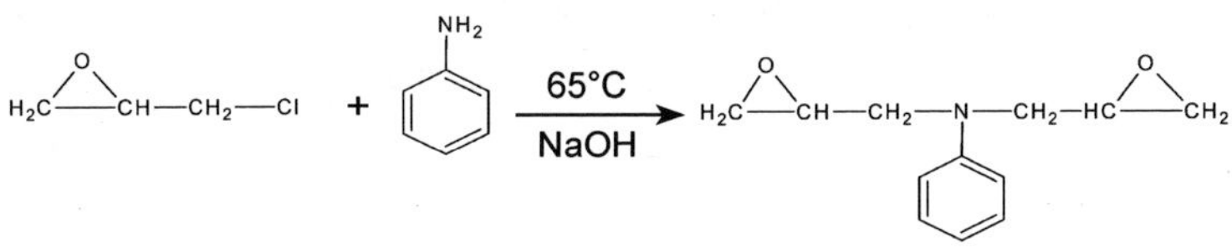

Epichlorohydrin Aniline Diglycidyl aniline

PROCEDURE

Prepare the azeotropic mixture of solvents by mixing 100 ml of methanol with 16 ml of water in a 500 ml three necked round bottom flask equipped with water condenser, thermometer and dropping funnel. Add 38 ml of aniline and 268 ml of epichlorohydrine to it. Flush the contents of the flask with oxygen free dry Nitrogen gas for about 10 min. and then reflux at 75°C for about 6-7 hours on water bath. Now add slowly and drop wise 100 ml of 40% aqueous sodium hydroxide solution over a period of 30 minutes at 60°C. The contents are further reflux for about 30 minutes at 60°C. After refluxing remove the excess of epichlorodrine and water by refluxing under vacuum. Then add 60 ml of benzene to the residue. Wash the solution repeatedly first with cold water and then with hot water by decantation. Vacuum distilled the organic benzene layer at 70°C to get the viscous aniline epoxy resin.

RESULT

Report the yield of Aniline epoxy resin

PROPERTIES

- These epoxy resins have remarkable chemical resistance and good adhesive properties.
- These have good dimensional stability and mechanical properties.
- When epoxy resins properly cured, yield very tough material.

APPLICATIONS

- These are excellent structural adhesives.
- These are used in industrial floorings, foams, potting materials for electrical insulations etc.
- These epoxy composites are used in aerospace and aircrafts.

❑❑❑

EXPERIMENT NO. 25

AIM

Preparation of Polyaniline.

APPARATUS

Three necked round bottom flask, Beaker, Funnel, Magnetic stirrer, Reflux condenser, Nitrogen gas cylinder, Thermometer etc.

CHEMICALS

Aniline , 0.1M Ammonium perdsulphate solution in 1M HCl, 1M Hydrochloric acid, Acetone, Distilled water.

REACTION

Polyaniline is a dark greenish blue colored conducting polymer and can be prepared by polymerization of aniline in presence of ammonium persulphate (initiator) in acidic condition at 0-2°C temperature. The conductivity of the product depends on the nature and amount of initiator and dopant used. The protonation, deprotonation and other various physico chemical properties of polyaniline are due to the presence of –NH group.

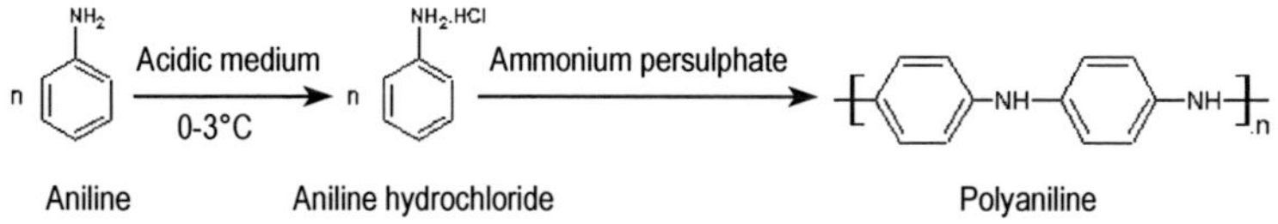

Polyaniline exists in three different interconvertible forms (Leucoemeraldine, emeraldine as salt or base and pernigraniline as salt or base forms) that differ in their chemical and physical properties. Out of these only the green protonated emaeraldine salt has the conductivity equivalent to semiconductors which is higher than that of the conductivity of common polymers but lower than that of the typical metals. The green protonated emaeraldine salt can be converted to non conducting deprotonated polyaniline (emaeraldine base) by treatment with alkali solution. The different forms of polyaniline and their interconversions are shown below.

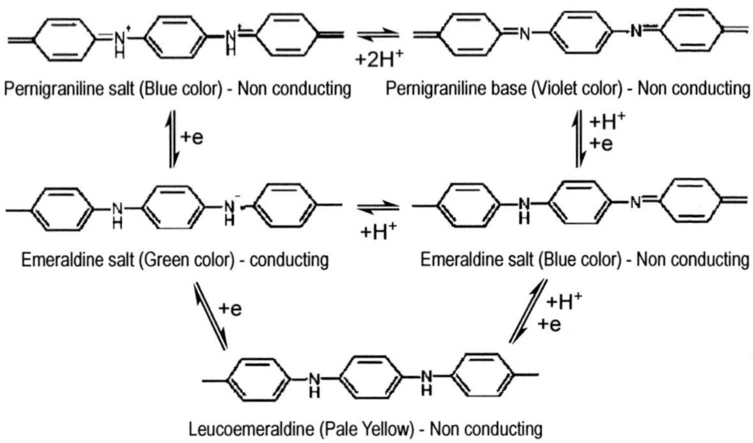

PROCEDURE

Take 5 ml of distilled aniline and 100 ml of 1 M hydrochloric acid solution in 250 ml conical flask. Mixed well and cool the content in ice bath below 3°C. Flush the content of conical flask with oxygen free nitrogen gas. Add 100 ml of 0.1M acidic ammonium persulphate solution from the burette slowly and drop wise to the conical flask containing acidic aniline solution with continuous stirring using magnetic stirrer for the period of about 30 minutes. During this addition the oxidation of aniline takes place which is an exothermic reaction, therefore the temperature of the reaction system increases. Hence maintain the temperature of the reaction system well below 3°C using ice bath during the course of addition. After complete addition, the reaction mixture is continuously stirred for about 6-7 hours. Polyaniline separates out as greenish blue colored precipitate. Filter the precipitate under suction, wash with 1M hydrochloric acid followed by water and finally with acetone. Dry it in an oven at 100°C. Take the weight of dry polymer.

RESULT

Report the yield of polyaniline

PROPERTIES

- It has desirable electrical, electrochemical and optical properties.
- It has excellent environmental stability.
- It is very inexpensive.

APPLICATIONS

- It is used in static films for transparent packaging of electronic components.
- It is used in anti-corrosion coatings for steels.
- It is used in electromagnetic shielding, rechargeable batteries, light emitting diodes, nonlinear optical devices, sensors for medical and pharmaceutical purposes etc.

❑ ❑ ❑

EXPERIMENT NO. 26

AIM

Preparation of styrene-acrylonitrile (SAN) copolymer.

APPARATUS

Three necked round bottom flask, Beaker, Funnel, Magnetic stirrer, Reflux condenser, Dropping funnel, Vacuum pump, Nitrogen gas cylinder, Vacuum oven, Thermometer etc.

CHEMICALS

Styrene, Acrylonitrile, Benzoyl peroxide (initiator), Benzene, Methanol, Hexanol, Distilled water.

REACTION

Styrene-acrylonitrile (SAN) copolymer can be prepared by the reaction between Styrene and Acrylonitrile in presence of benzoyl peroxide as an initiator.

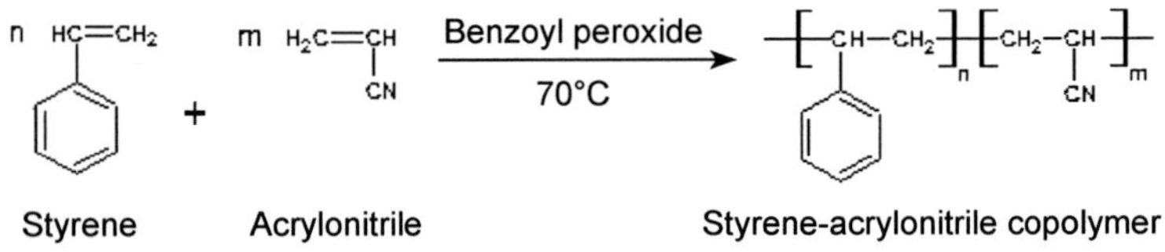

PROCEDURE

Take 100 mg of benzoyl peroxide in a 250 ml three necked round bottom flask equipped with water condenser, thermometer and dropping funnel. Add 10 ml each of distilled Styrene and Acrylonitrile to the flask and shake gently to dissolve the initiator in it. Flush the contents of the flask with oxygen free dry Nitrogen gas for about 10 min. Reflux the content of the flask on water bath at 70°C. The beginning of polymerization is indicated by the increase in the viscosity of the content of the flask. Stop refluxing the content at proper time and allow it to cool. Dilute the mixture with benzene or chloroform carefully if necessary to obtain the fluidity. Transfer the content of the flask drop wise in 300 ml of methanol or hexane or petroleum ether. Collect the precipitate on frilled glass funnel and dry in a vacuum oven at 80°C. Take the weight of dry copolymer.

RESULT

Report the yield of styrene-acrylonitrile (SAN) copolymer

PROPERTIES

- It is transparent and brittle in nature.
- It has very good impact strength and chemical resistance.
- It has good thermal resistance than that of polystyrene polymer

APPLICATIONS

- It is used for moulding crockery items and machine components.
- It is widely used in place of polystyrene owing to its greater thermal resistance.

❑❑❑

EXPERIMENT NO. 27

AIM

Preparation of styrene-methylmethacrylate copolymer.

APPARATUS

Three necked round bottom flask, Beaker, Funnel, Magnetic stirrer, Reflux condenser, Dropping funnel, Vacuum pump, Nitrogen gas cylinder, Vacuum oven, Thermometer etc.

CHEMICALS

Styrene, Methylmethacrylate, Benzoyl peroxide (initiator), Toluene, Petroleum ether, Distilled water.

REACTION

Styrene-methylmethacrylate copolymer can be prepared by the reaction between Styrene and methylmethacrylate in presence of benzoyl peroxide as an initiator.

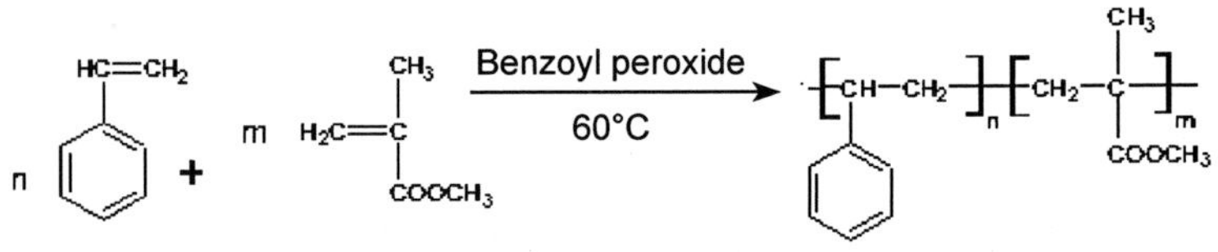

Styrene Mehylmethacrylate Styrene-mehylmethacrylate copolymer

PROCEDURE

Take 50 mg of benzoyl peroxide in a 250 ml three necked round bottom flask equipped with water condenser, thermometer and dropping funnel. Add 10 ml each of distilled Styrene and methylmethacrylate to the flask and shake gently to dissolve the initiator in it. Flush the contents of the flask with oxygen free dry Nitrogen gas for about 10 min. Reflux the content of the flask on water bath at 60°C. The beginning of polymerization is indicated by the increase in the viscosity of the content of the flask. Stop refluxing the content at proper time and allow it to cool. Dilute the mixture with toluene carefully if necessary to obtain the fluidity. Transfer the content of the flask drop wise in 300 ml of petroleum ether. Collect the precipitate on frilled glass funnel and dry in a vacuum oven at 60°C. Take the weight of dry copolymer.

RESULT

Report the yield of styrene-methylmethacrylate resin

PROPERTIES

- It is used as an alternative to acrylic resins.
- It has low density.
- It has low moisture absorption.
- It has good moldability.

APPLICATIONS

- It is used for flat display substrates.
- It is used for optical components.
- It is used for manufacturing miscellaneous goods.

❑❑❑

SECTION— III

5. Experimental Characterization of Polymers

5

Experimental Characterization of Polymers

The various chemical and analytical techniques are available for the characterization of the polymers. This chapter describes some of the important but simple laboratory experimental procedure for the characterization of common polymers. The principles involved in the chemical and analytical techniques used for the characterization of common polymers are already explained in chapter 3 of section-I.

EXPERIMENT NO. 1

AIM

Determination of Percentage moisture content of given polymer sample.

APPARATUS

Digital electronics balance (with accuracy 0.1mg), porcelain crucible, Oven, Desiccator.

CHEMICALS

Polymer sample.

THEORY

The moisture content is the characteristics of plastics depend largely upon the basic type and composition of the material.

Knowledge of the moisture content is important in understanding performance, processing and stability since some polymers degrade when molded, resulting in reduced properties. The heating time and temperature values can be arrived by testing a same material by varying the temperature and heating time independently to determine optimal conditions for removal of all water in the sample. If the temperature is too low or the heating time too short all of the water might not be driven off. A temperature value too high can degrade the polymer and actually generate water in some cases. There are different methods for determinations of moisture content of polymer sample but the gravimetric method is the simplest one for quantitative determination of moisture content of polymer.

The moisture content of polymer can be calculated by using the following equation,

$$\text{Percentage moisture content} = \frac{(W_2 - W_3)}{(W_2 - W_1)} \times 100$$

Where W_1 = Weight of empty porcelain crucible

W_2 = Weight of porcelain crucible with sample

W_3 = Constant weight of porcelain crucible with sample after heating.

PROCEDURE

Take a weight of clean and dried porcelain crucible (W_1). Then place small quantity of polymer sample (approximately 2g) in it and take the weight again (W_2). Now keep the crucible in an oven maintained at 100°C for about 2 hours. After the completion of heating of sample, the crucible is to be taken out of an oven and allow it to cool to room temperature and then place it in desiccator for some time. Take the weight of crucible (W_3). Repeat the same procedure till to get the constant weight W_3. Typically the experiment is perform three times with equal quantity of sample and the average result is reported.

OBSERVATION

1. Weight of empty porcelain crucible (W_1) =
2. Weight of porcelain crucible with sample (W_2) =

3. Weight of porcelain crucible with sample after heating (W_3) =
4. Weight of sample taken = $W_2 - W_1$
5. Weight loss of polymer after heating = $W_2 - W_3$

CALCULATION

The percentage moisture content of the polymer sample can be calculated by using the equation,

$$\text{Percentage moisture content} = \frac{(W_2 - W_3)}{(W_2 - W_1)} \times 100$$

RESULT

The percentage moisture content of the given polymer sample is found to be = ... %.

❑❑❑

EXPERIMENT NO. 2

AIM

Determination of ash content of given vulcanized rubber sample.

APPARATUS

Digital electronics balance (with accuracy 0.1mg), silica crucible, Muffle furnace, Desiccator.

CHEMICALS

Vulcanized rubber sample.

THEORY

Ash content of the rubber is due to the inorganic fillers in the rubber sample. Thus this test will identify the total filler content of the polymer. It cannot identify individual percentages in multi-filled materials without additional test procedures being performed. An ash test cannot be used to determine the percent carbon fiber or percent carbon black since carbon burns off during the ash test. An Ash test involves taking a known amount of sample, placing the weighed sample into a dried / pre-weighed silica crucible, burning away the polymer in an air atmosphere at temperatures above 500°C, and weighing the crucible after it is has been cooled to room temperature in a desiccator. Ash residue remaining in the crucible is considered filler unless the residue is less than 1%. Residues of more than 1% are typically the result of additives that did not burn off.

The percentage ash content of the rubber sample can be calculated by using the following equation,

$$\text{Percentage ash content} = \frac{(W_3 - W_1)}{(W_2 - W_1)} * 100$$

Where W_1 = Weight of empty silica crucible

W_2 = Weight of silica crucible with sample

W_3 = Weight of silica crucible with sample after pyrolysis.

PROCEDURE

Take a weight of clean and dried silica crucible (W_1). Then place small quantity of polymer sample (approximately 2g) in it and take the weight again (W_2). Now keep the crucible in muffle furnace maintained at 600-800°C for about 4-5 hours. After the completion of pyrolysis of rubber, the crucible is to be taken out of furnace and allow it to cool to room temperature and then place in desiccator for some time. Take the weight of crucible (W_3). Typically the experiment is perform three times with equal quantity of sample and the average result is reported.

OBSERVATION

1. Weight of empty silica crucible (W_1) =
2. Weight of silica crucible with sample (W_2) =
3. Weight of silica crucible with sample after pyrolysis (W_3) =
4. Weight of sample taken = $W_2 - W_1$
5. Weight of residue = $W_3 - W_1$

CALCULATION

The percentage ash content of the rubber sample can be calculated by using the equation,

$$\text{Percentage ash content} = \frac{(W_3 - W_1)}{(W_2 - W_1)} * 100$$

RESULT

The percentage ash content of the given rubber sample is found to be = %.

❑❑❑

EXPERIMENT NO. 3

AIM

Determination of density or Specific gravity of powdered Polymer sample.

APPARATUS

Digital electronic balance (with accuracy 0.0001g), Density bottle etc.

CHEMICALS

Given powdered polymer sample, Non-solvent for the given polymer sample, acetone, distilled water.

THEORY

The flotation method for the density measurement is suitable for polymer samples which are in the form of beads and cannot be applicable for powdered form of polymer sample. In case of powdered polymer sample the density can be determined by the method as explained in the procedure and using following equation,

Specific gravity

$$= \frac{(W_4 - W_1) * \dfrac{(W_3 - W_1)}{(W_2 - W_1)}}{(W_4 - W_1) + (W_3 - W_5)}$$

Where W_1 = Weight of empty density bottle

W_2 = Weight of density bottle with water

W_3 = Weight of density bottle with non-solvent

W_4 = Weight of density bottle filled three-fourth with powdered polymer

W_5 = Weight of density bottle filled three-fourth with powdered polymer and one-fourth with non-solvent

PROCEDURE

Take the constant weight of clean and dried empty density bottle (W_1). Fill it with distilled water and take the weight (W_2). Remove the water, rinse the density bottle with acetone followed by non-solvent and then fill it with the same non-solvent and take the weight (W_3). Now again fill the same empty dried density bottle up to its three-fourth volume with given powdered polymer sample and take the weight (W_4). Then after weighing fill the remaining one-fourth volume of the density bottle with non-solvent and take the weight (W_5). The specific gravity or density of the powdered polymer sample can be calculated by using the equation given above.

OBSERVATIONS

1. Weight of empty density bottle ($W_{1)}$ =
2. Weight of density bottle with water (W_2) =
3. Weight of density bottle with non-solvent (W_3) =
4. Weight of density bottle filled three-fourth with powdered polymer (W_4) =
5. Weight of density bottle filled three-fourth with powdered polymer and one-fourth with non-solvent (W_5) =

CALCULATIONS

Calculate the density by using the equation,

Specific gravity

$$= \frac{(W_4 - W_1) * \dfrac{(W_3 - W_1)}{(W_2 - W_1)}}{(W_4 - W_1) + (W_3 - W_5)}$$

RESULT

The density of given powdered polymer sample is found to be ---- g/cc.

❑❑❑

EXPERIMENT NO. 4

AIM

Determination of density or Specific gravity of Polymer by displacement method.

APPARATUS

Digital electronic balance (with accuracy 0.0001g), Density bottle etc.

CHEMICALS

Given polymer sample, Nylon threads having negligible weight, Sinker, distilled water.

THEORY

Density is the ratio of mass per unit volume at a specific temperature and expressed as g/cm^3 or kg/m^3. Displacement method involves the weighing of the polymer sample in an air and in water. If the sample is having density less than water then the weight of sample in water is to be taken along with sinker. The density can be determine by using following equation,

$$\text{Density} = \frac{\text{Weight of sample in air}}{\text{Loss in weight of sample in water}} \times \text{Density of water}$$

In case of polymer sample having density more than water, the density can be determined by using following equation,

$$\text{Density} = \frac{W_1}{(W_1 - W_2)} \times \text{Density of water}$$

Where W_1 = Weight of sample in air

W_2 = Weight of sample in water

In case of polymer sample having density less than water, the density can be determined by sinker method using following equation,

$$\text{Density} = \frac{W_1}{W_1 - (W_4 - W_3)} \times \text{Density of water}$$

$$\text{Density} = \frac{W_1}{(W_1 + W_3 - W_4)} \times \text{Density of water}$$

Where W_1 = Weight of sample in air

W_3 = Weight of sinker in water

W_4 = Weight of sample along with sinker in water

PROCEDURE

Take a weight of polymer sample (W_1) having smooth surface and free from dust. If the sample having density more than water i.e. dips in water, note down the weight of sample in distilled water (W_2) by tying the sample with weightless thin nylon thread. However if the sample having density less than water i.e. floats on water, take the weight of sample along with the sinker in water (W_4) by tying the sample and sinker with weightless thin nylon thread. Also record the weight of sinker alone in water (W_3) by tying it with weightless thin nylon thread.

OBSERVATIONS

1. Weight of polymer sample (W_1) =
2. Weight of polymer sample in water (W_2) =
3. Weight of sinker in water (W_3) =
4. Weight of polymer sample along with sinker in water (W_4) =
5. Loss in weight of polymer sample in water ($W_1 - W_2$) =

   OR

   Loss in weight of polymer sample in water with sinker [$(W_4 - W_3)$] =

CALCULATIONS

Calculate the density of polymer sample by using one of the following applicable equation,

In case of polymer sample having density more than water,

$$\text{Density} = \frac{W_1}{(W_1 - W_2)} \times \text{Density of water}$$

In case of polymer sample having density less than water,

$$\text{Density} = \frac{W_1}{(W_1 + W_3 - W_4)} \times \text{Density of water}$$

RESULT

The density of given polymer sample by displacement method is found to be — g/cc.

❑❑❑

EXPERIMENT NO. 5

AIM

Determination of density of polymer (Polystyrene) by flotation method.

APPARATUS

Measuring cylinder, Density bottle, Burette, Beaker, etc.

CHEMICALS

Given Polystyrene sample, Sodium chloride salt, water.

THEORY

The flotation method for the density measurement is suitable for polymer beads sample. The method involves the addition of higher density liquid to the liquid of lower density containing the polymer sample until the test polymer sample particles neither sink nor rise to the surface that is the particles should float exactly at the central place in the bulk of total liquid. At this situation the density of the test polymer and the liquid mixture or solution are equal and hence the density of the polymer sample can be known from the density measurement of the liquid mixture or solution.

For the determination of the density of polymers having density less than one (*e.g.* Polyethylene), Ethanol-Water mixture is suitable, however for the polymers having density more than one (*e.g.* Polystyrene) one may use the solution of water and water soluble salt.

PROCEDURE

Fill up graduated measuring cylinder up to the 100 ml mark with water. Place a crystal of polymer sample polystyrene in it. As the density of polystyrene is more than that of water it get settle down at the bottom of water. The density of water then can be increased by adding solid Sodium chloride salt in water with continuous stirring and the position of polymer sample is observed in solution. The process of addition of salt to water is continued till the polymer crystal starts floating exactly at the centre of the solution. At this stage the density of solution is measured using density bottle or pyknometer which is equal to the density of the polystyrene sample.

While adjusting position of polystyrene crystal at the centre of the solution, if the crystal starts floating in the upper part of solution due to excess addition of salt then water can be added to it to bring it at the centre.

OBSERVATIONS

1. Volume of density bottle (V) =
2. Weight of empty density bottle (W_1) =
3. Weight of density bottle filled with solution (W_2) =
4. Weight of solution ($W_2 - W_1$) =

CALCULATIONS

Calculate the density of solution by using the following equation,

$$\text{Density of solution} = \frac{(W_2 - W_1)}{V}$$

This density of solution is equal to the density of polymer sample.

RESULT

The density of Polystyrene by flotation method is found to be---- g/cc

❑❑❑

EXPERIMENT NO. 6

AIM

Separation and purification of polymer (Polymethyl methacrylate).

APPARATUS

Beaker, Glass rod, Magnetic stirrer, Suction pump, Sintered glass crucible.

CHEMICALS

Polymethyl methacrylate, Chloroform, Methanol.

THEORY

Polymers are always synthesized in the laboratory by solution or emulsion polymerization methods. Separation methods of polymers from the reaction mixture depend on the properties of the polymer and the system of polymerization. Polymers prepared in solution are usually isolated by adding the reaction mixture in excess of non-solvent which cause the polymer to precipitate out and can be filtered or strained. Sometimes the precipitated polymer remains in the form of suspended colloid, under such condition polymer can be removed by cooling the mixture with ice or by adding an electrolyte like sodium chloride or diluted acid solution. Other methods such as long stirring time, shaking or centrifuging can also complete the separation.

Separation of polymers from aqueous emulsion is often difficult to carry out. Methods frequently used are diluting using aqueous electrolyte solution, freezing or pouring or spraying into alcohol.

Polymers separated using methods described above are likely to get contaminated by impurities such as residual monomer, initiator, solvent and possibly additives like surfactant or transfer agent. Purification of the polymer using methods such as distillation and recrystallisation is obviously impossible to carry out. Instead methods like re-precipitation, freeze-dry, normal drying and solvent extraction are used.

PROCEDURE

(**Note :** *Students will be provided with 10ml solution of poly (methyl methacrylate) in chloroform having a concentration of about 5% w/v. The polymer is separated and purified using the re-precipitation method as explained below).*

Separate the polymer from the polymer solution by precipitating it in non-solvent methanol. Add the entire polymer solution drop wise into the methanol with a volume of more than 5 times the volume of the polymer solution. (~50 ml). While adding the polymer solution, the mixture must be stirred vigorously either with a magnetic stirrer or glass rod. When precipitation is completed, filter the polymer using a clean and dried sintered glass with its mass known beforehand. Wash the polymer with a small amount of methanol and leave it to dry by suction (vacuum pump) for about 20 min. Dry the sample in vacuum oven at about 50°C at least for a night. After drying, weigh the container again and record the weight of the polymer product.

RESULT

The weight of purified polymer is found to be = ...

Note: Experiment can also be performed by selecting the following pair of solvent and non-solvent system for the given polymer.

Sr. No.	Polymer	Solvent	Non-Solvent
1.	Polystyrene	Toulene/ Benzene	Methanol
2.	Polyethylene glycol	Water (Cold)	Ether
3.	Nylon-66	Formic acid	Water
4.	Polyvinyl alcohol	Water	Acetone
5.	Polyvinyl chloride	THF	Methanol

❑❑❑

EXPERIMENT NO. 7

AIM

Determination of the acid value of given plastic material.

APPARATUS

Burette, pipette, conical flask, glass rod etc.

CHEMICALS

Ethyl alcohol, Std. 0.1 N KOH solution, Plastic, Phenolphthalein, Acetone, Benzene etc.

THEORY

The acid value is number of milligrams of KOH which are needed to neutralize the acidity present in 1 g of material. For its determination, material is dissolved in suitable solvent usually neutral ethyl alcohol. It is then titrated against standard alkali solution (0.1 N KOH) using phenolphthalein as an indicator. Acid value is calculated by formula,

$$\text{Acid Value} = \frac{\text{Volume of KOH} * \text{Normality of KOH} * 56}{\text{Weight of sample taken}}$$

PROCEDURE

Standardized the KOH solution prepared by titrating it with standard oxalic acid solution and find the exact concentration of KOH solution. Take exactly 5 g of the given plastic sample in round bottom flask and Add 50 ml of neutral ethanol to it. Reflux the solution till to get the clear solution using water bath then cool it. Add 2 to 3 drops of phenolphthalein indicator in it. Titrate the solution with 0.1 N KOH till the color of solution changes to pale pink. If the sample is insoluble in ethanol, acetone or benzene can be used to dissolved sample and the titration can be performed with 0.1N ethanolic KOH.

OBSERVATION

1. Normality of standardized KOH solution = N_{KOH}
2. Volume of KOH solution required for titration = V_{KOH} ml
3. Weight of sample taken = W g

CALCULATIONS

1000 ml of 1 N KOH = 56 g of KOH

V ml of N Normal KOH = (W g) of KOH

$$W \text{ g of KOH} = \frac{V * N * 56}{1000}$$

This W g of KOH = acid value of dissolved weight of sample

Hence acid Value in g of KOH required per g of sample

$$= \frac{V * N * 56}{\text{Weight of sample taken} * 1000}$$

Hence acid Value in mg of KOH required per g of sample

$$= \frac{V * N * 56}{\text{Weight of sample taken}}$$

RESULT

Acid value of given plastic material is... mg of KOH.

Note: Following samples with their expected acid values can be provided for the determination of acid value.

Sr. No.	Name of plastic	Acid value (mg of KOH / g of plastic)
1.	Methyl cellulose	< 3
2.	Benzyl cellulose	< 3
3.	Polyvinyl acetatals	< 5
4.	Polystyrene	< 5
5.	Polyisobutylene	5
6.	Polyvinyl alcohol	< 10
7.	Polyvinyl acetate	< 10
8.	Ethyl cellulose	< 10
9.	Phenol formaldehyde resin	< 20
10.	Unsaturated polyesters	< 50

❑❑❑

EXPERIMENT NO. 8

AIM

Determination of carbonyl value of given polymer sample.

APPARATUS

Electronics balance, Iodine flask, Water condenser, Stopper, Burette, Pipette, etc.

CHEMICALS

Given polymer sample, 0.5N Ethanolic Potassium hydroxide solution, 0.2N Hydrochloric acid, Hydroxyl amine solution, Ethanol, Bromophenol Blue indicator.

THEORY

The Carbonyl value of the polymer is the number of milligrams of KOH equivalent to the hydroxylamine necessary to oximate 1 g of sample.

Carbonyl value of given polymer sample can be determined by using the following equation

$$\text{Carbonyl value} = \frac{11.22 * (X - Y)}{W} - \text{Acid value}$$

Where W = Weight of sample in g

X = Volume of 0.2N HCl solution required for Blank titration

Y = Volume of 0.2N HCl solution required for Experimental titration

PREPARATION OF SOLUTIONS

1. **Hydroxylamine Solution**

   Dissolved 4.0 g of Hydroxyl amine hydrochloride in 10ml of distilled water and then add 80ml of ethanol.

The solution is then treated with 60ml of 0.5N Ethanolic Potassium hydroxide solution with continuous stirring.

2. **Bromophenol Blue indicator solution**

   Dissolved 0.4 g of Bromophenol Blue in 12 ml of 0.05N sodium hydroxide and make up the volume to 100ml with distilled water.

PROCEDURE

The experiment is performed in two parts:

PART I: Determination of acid value of the given polymer sample.

Determine the acid value of the given polymer sample as explained in previous experiment (Expt. No. 7).

PART II: Treatment of sample with hydroxylamine solution.

Take accurately weighted 1 g of given sample in a conical flask and dissolved it in a little amount of ethanol or any other water miscible carbonyl free solvent. It is then mixed with 50 ml of hydroxylamine solution. Add 1 ml of Bromophenol blue indicator solution. Cool the flask and titrate the excess of KOH with 0.2N HCl until the colour changes from purple to yellow (Experimental Y ml). Perform the blank experiment by repeating the same procedure but without taking the sample (X ml).

OBSERVATIONS

1. Weight of empty weighing bottle = W_1 g
2. Weight of weighing bottle with sample = W_2 g
3. Weight of sample taken = W = $(W_2 - W_1)$ g
4. Volume of 0.2N HCl solution required for Experimental titration = Y ml
5. Volume of 0.2N HCl solution required for Blank titration = X ml

CALCULATIONS

Determine the carbonyl value of the given sample by using the following equation,

$$\text{Carbonyl value} = \frac{11.22 * (X - Y)}{W} - \text{Acid value}$$

RESULT

The carbonyl value of the given polymer sample is found to be = …

❑❑❑

EXPERIMENT NO. 9

AIM

Determination of Hydroxyl value of a resin.

APPARATUS

Water bath, water condenser, conical flask, burette, pipette etc.

CHEMICALS

Acetic anhydride, Pyridine, 1M KOH solution, Phenolphthalein indicator.

THEORY

Total basicity, base value or hydroxyl value or base number or hydroxyl number of a polymer is the amount of titrating acid, expressed as mg of equivalent KOH, required to neutralize all the basic constituents of 1gm of the sample. Hydroxyl value is measured by end group analysis of hydroxyl group containing polymer resin by acetylating polymer with acetic anhydride in pyridine with respect to blank and resulting mixtures are then titrated with standardized alkali (KOH) solution. If X and Y are the volume of KOH required for Blank and Experimental titration of mixtures respectively, then the number of hydroxyl groups can the determine by using the following mathematical calculations,

1000 ml of 1 N KOH = 56.1 g of KOH

(X – Y) ml of N Normal KOH = (W g) of KOH

$$\text{W g of KOH} = \frac{N * 56.1 * (X - Y)}{1000}$$

This W g of KOH = Hydroxyl value of dissolved weight of sample,

Hence Hydroxyl Value in g of KOH required per g of sample

$$= \frac{56.1 * \text{N of KOH} * (X - Y)}{\text{Weight of polymer sample taken} * 1000}$$

Number of Hydroxyl groups in mg of KOH

$$= \frac{56.1 * \text{N of KOH} * (X - Y)}{\text{Weight of polymer sample taken}}$$

PROCEDURE

Weigh accurately 0.25 g of polymer sample and transfer in 250 ml conical or Round Bottom flask. Add 5ml of Acetic anhydride and 10ml of pyridine. Put porcelain pieces in the flask and reflux the mixture in water bath using water condenser. Continue the refluxing for half hour. The flask allow to cool to room temperature and then add 70ml of water from the side of the flask, shake well the mixture and titrate the complete mixture against 1N KOH solution using Phenolphthalein indicator (EXPERIMENTAL). Repeat the same procedure without taking the polymer sample (BLANK). The end point is detected by the colour change from colourless to pink.

REACTIONS

$$\text{Polymer-OH} + Ac_2O \rightarrow \text{Polymer-OAC} + ACOH$$

$$Ac_2O + H_2O \rightarrow 2ACOH$$

$$ACOH + KOH \rightarrow ACOK + H_2O$$

OBSERVATIONS

1. Standardized 1N KOH solution by titrating it against standard oxalic acid solution and find out the accurate concentration of KOH solution used (N_{KOH}).
2. Burette reading for Experimental titration: Y ml of KOH
3. Burette reading for Blank titration: X ml of KOH

CALCULATIONS

Determine the number of hydroxyl groups *i.e.* hydroxyl value by using the following equation,

Number of Hydroxyl groups in mg of KOH

$$= \frac{56.1 * \text{N of KOH} * (X - Y)}{\text{Weight of polymer sample taken}}$$

RESULT

Hydroxyl value in mg of KOH of given resin is found to be = ...

Note: Following samples with their expected hydroxyl values can be provided for the determination of hydroxyl value.

Sr. No.	Name of plastic	Hydroxyl value (mg of KOH / g of plastic)
1	Urea formaldehyde	0
2	Melamine formaldehyde	0
3	Polyvinyl chloride	0
4	Polystyrene	0
5	Polyesters	0-350
6	Phenol formaldehyde resin	125-450
7	Cellulose nitrate	20-250
8	Polyvinyl acetatals	65-250
9	Polyvinyl alcohol	1000-1270

❑❑❑

EXPERIMENT NO. 10

AIM

Determination of the Iodine value of given plastic material.

APPARATUS

Burette, pipette, 250 ml Iodine flask, glass rod etc.

CHEMICALS

0.1 N Sodium thiosulphate solution, Plastic material, Starch indicator, Carbon tetrachloride, 0.1M Iodine monobromide solution, Potassium iodide etc.

THEORY

The Iodine value is a measure of degree of unsaturation of the material. The addition of iodine to unsaturated linkages, especially carbon-carbon double bond, triple bond can be used for iodine value estimation. Thus it can be defined as the number of grams of iodine that reacts with 100 g of material. The iodine value can be experimentally determined by titrating unreacted iodine with respect to blank against sodium thiosulphate solution. If X and Y are the volume of Sodium thiosulphate solution required for Blank and Experimental titration of mixtures respectively, then the iodine value can be calculate by using the following mathematical calculations,

1000 ml of 1 N $Na_2 S_2O_3$ solution = 126.9 g of I_2

(X – Y) ml of N Normal $Na_2 S_2O_3$ = How much grams of I_2

$$\text{grams of } I_2 = \frac{(X - Y) * \text{Normality of } Na_2 S_2O_3 * 126.9}{1000}$$

grams of I_2 per g of sample

$$= \frac{(X - Y) * \text{Normality of } Na_2 S_2O_3 * 126.9}{\text{Weight of sample taken} * 1000}$$

grams of I_2 per 100 g of sample

$$= \frac{(X - Y) * \text{Normality of } Na_2\,S_2O_3 * 126.9 * 100}{\text{Weight of sample taken} * 1000}$$

$$\text{Iodine Value} = \frac{(X - Y) * \text{Normality of } Na_2\,S_2O_3 * 126.9}{\text{Weight of sample taken (W)} * 10}$$

PROCEDURE

Take exactly 0.1 to 1 g depending on the iodine value of the given plastic material in 250 ml iodine flask and dissolve it in 25 ml of carbon tetrachloride and 20 ml of 0.1M iodine monobromide solution. Seal the flask, mixed well and stand for about one hour. Add 10 ml of 20% aqueous potassium iodide solution and 100 ml of distilled water. Titrate the complete mixture with 0.1N sodium thiosulphate solution using starch as an indicator till to get the end point from blue to colorless. This gives experimental titration value (Y). Repeat the same procedure without taking the sample which gives blank titration value (X).

Note: Quantity of sample to be taken depending on the iodine value of the sample is given in following table.

I_2-Value	Quantity of sample (g)
Less than 20	1.0
20 – 60	0.5 – 0.25
60 – 100	0.25 – 0.15
More than 100	0.15 – 0.10

OBSERVATION

1. Weight of sample taken = W g
2. Normality of $Na_2\,S_2O_3$ solution = 0.1 N
3. Volume of $Na_2\,S_2O_3$ solution required for Blank titration = X ml
4. Volume of $Na_2\,S_2O_3$ solution required for Experimental titration = Y ml

CALCULATION

The iodine value of the given polymer sample can be calculated by using the following equation,

$$\text{Iodine Value} = \frac{(X - Y) * \text{Normality of } Na_2S_2O_3 * 126.9}{\text{Weight of sample taken} * 10}$$

RESULT

Iodine value of given plastic material is found to be = ...

Note: Following samples with their expected iodine values can be provided for the determination of iodine value.

Sr. No.	Name of plastic	Iodine value
1	Polyisobutylene	1
2	Butyl rubber	5
3	Natural rubber	345-375
4	Polybutadiene	385-440

❑❑❑

EXPERIMENT NO. 11

AIM

Determination of saponification value of a polymer.

APPARATUS

Water bath, water condenser, conical flask, burette, pipette etc.

CHEMICALS

0.5N KOH in ethanol solution, 0.5N Hydrochloric acid, Phenolphthalein indicator, Methanol, given polymer sample.

THEORY

Saponification value is the measure of saponifiable matter present in the sample. It can be defined as number of milligram of potassium hydroxide required for complete saponification of one gram of sample. As the amount of alkali required to saponify amount of polymer sample depends on the molecular weight of the fatty acid present in the sample, saponification value tells us about the proportion of the fatty acid present in the sample. The experimental determination of Saponification value of a polymer involves the saponification that is alkaline hydrolysis of ester followed by tration of unreacted alkali against standardized Hydrochloric acid. If X and Y are the volume of HCl required for titration of unreacted KOH of Blank and Experimental titration of mixtures respectively, then the saponification value can the determine by using the following mathematical calculations,

1000 ml of 1 N HCl = 56.1 g of KOH

(X – Y) ml of N Normal HCL = How much g of KOH

$$\text{g of KOH} = \frac{(X - Y) * \text{Normality of HCl} * 56.1}{1000}$$

$$\text{mg of KOH} = \frac{(X - Y) * \text{Normality of HCl} * 56.1}{1000} * 1000$$

mg of KOH = (X – Y) * Normality of HCl * 56.1

mg of KOH required per g of sample *i.e.* saponication value

$$\text{Saponication value} = \frac{(X - Y) * \text{Normality of HCl} * 56.1}{\text{Weight of polymer sample taken (W)}}$$

PROCEDURE

Standardized 0.5N KOH solution by titrating it against standard oxalic acid solution and find out the accurate concentration of KOH solution (N_{KOH}). Then standardized 0.5N HCl solution by titrating it against above standardized 0.5N KOH solution and find out the accurate concentration of HCl solution (N_{HCl}).

Weigh accurately 1 g of polymer sample and transfer in 250 ml conical. Add 50ml of ethanolic Potassium hydroxide solution. Put porcelain pieces in the flask and reflux the mixture using water condenser for one hour. The flask is allowed to cool and titrate the excess of alkali in warm mixture against 0.5N HCl solution using Phenolphthalein indicator (EXPERIMENTAL). The end point is detected by the colour change from pink to colourless. Repeat the same procedure without taking the polymer sample (BLANK).

OBSERVATIONS

1. Weight of polymer sample taken = W g
2. Normality of standardized KOH solution used = N_{KOH}
3. Normality of Standardized HCl solution used = N_{HCl}
4. Volume of HCL solution required for Experimental titration = Y ml
5. Volume of HCL solution required for Blank titration = X ml

CALCULATIONS

Determine the saponification value by using the following equation,

$$\text{Saponication value} = \frac{(X - Y) * \text{Normality of HCl} * 56.1}{\text{Weight of polymer sample taken (W)}}$$

RESULT

The Saponification value of the given polymer is found to be =...

Note: Following samples with their expected Saponification values can be provided for the determination of saponification value.

Sr. No.	Name of plastic	Saponification value (mg of KOH / g of plastic)
1.	Urea formaldehyde resin	< 20
2.	Pheol formaldehyde resin	< 20
3.	Polymethyl methacrylate	< 20
4.	Polyethylene	< 20
5.	Polystyrene	< 20
6.	Natural rubber	< 20
7.	Butyl rubber	< 20
8.	Polyninyl chloride	< 20
9.	Polyvinyl acetates	100-650
10.	Cellulose acetates	>200
11.	Polyacrylic acid	>500
12.	Polyurethane	>500

❑❑❑

EXPERIMENT NO. 12

AIM

Determination of Acetyl Content in a given Polymer sample (Cellulose Acetate).

APPARATUS

250ml Conical or Round Bottom flask with joint, Reflux Condenser, Water Bath, Pipette, Burette.

CHEMICALS

Given Polymer sample (Cellulose Acetate), Pyridine, 0.5N KOH solution, 0.5N HCL solution, Phenolphthalein Indicator

PROCEDURE

Weigh accurately 0.5 to 1g of given polymer sample and transfer quantitatively in 250 ml conical or Round Bottom flask. Add 25ml of freshly distilled Pyridine to this flask and shake well till the polymer gets completely dissolved. Then add the 20ml of 0.5N KOH solution to this polymer solution and reflux the mixture in water bath using water condenser. Put porcelain pieces in the flask to avoid bumping of the solution. Continue the refluxing for one and half hour. The flask is allowed to cool to room temperature and then titrate the complete mixture against 0.5N HCL solution using Phenolphthalein indicator.

CALCULATIONS

1. Standardized 0.5N KOH solution by titrating it against standard oxalic acid solution and find out the accurate concentration of KOH solution used (N_{KOH}).
2. Standardized 0.5N HCL solution by titrating it against above Standardized KOH solution and find out the accurate concentration of HCL solution used (N_{HCL}).

   Perform the calculations as given below,

   Accurate weight of polymer sample taken = W g

Amount of KOH (N_{KOH}) taken= 20 ml

Amount of HCL (N_{HCL}) required for titration= X ml

Amount of KOH (N_{KOH}) consumed= (20-X) ml= Yml

Acetyl content (%) =

$$\frac{Y * N \text{ of KOH} * \text{Mole. wt. of } CH_3CO \text{ group} * 100}{1000 * \text{Weight of polymer sample taken}} = Z_1\ \%$$

Similarly acetic acid content can determine by using following equation,

Acetic acid content (%) =

$$\frac{Y * N \text{ of KOH} * \text{Mole. wt. of } CH_3COOH \text{ group} * 100}{1000 * \text{Weight of polymer sample taken}} = Z_2\ \%$$

OR

$$\text{Acetic acid content } (Z_2\%) = \frac{Z_1 * \text{Mole. wt. of } CH_3COOH \text{ group}}{\text{Mole. wt. of } CH_3CO \text{ group}}$$

RESULT

The acetyl content of given polymer sample (Z_1 %) =

Acetic acid content of given polymer sample (Z_2 %) =

❑❑❑

EXPERIMENT NO. 13

AIM

Determination of Epoxy Equivalent Weight (EEW) of an Epoxy resin by hydrochloric acid-acetone-standing method.

APPARATUS

Conical flask, Stopper, Burette, Pipette etc.

CHEMICALS

Given Epoxy resin, Hydrogen Chloride-Acetone mixture, Acetic anhydride, Pyridine, 0.1N Sodium Hydroxide solution, Phenolphthalein indicator.

THEORY

Epoxy equivalent weight (EEW) is the most important features of epoxy resin. By knowing EEW, the required amount of the cross-linking agent may be calculated. Epoxy resins are high molecular weight polymers which normally contain at least two epoxide groups. These epoxide groups are very active and may have simple reaction with various cross-linking agents, such as amines, hydroxyls, carboxyls etc. to produce materials with different properties. Therefore, it is crucial to control or to know the epoxide content during production of epoxy resins or formulation of a specific product involving epoxy resins in which various cross-linking agents in stoichiometric proportions are required. Epoxide content is commonly expressed as the epoxide equivalent weight (EEW), which is defined as the weight in grams of an epoxy resin containing one mole equivalent of epoxide (g/mol), *i.e.* one gram-equivalent of epoxy.

$$\text{EEW} = \frac{\text{Weight of polymer molecule}}{\text{number of epoxy groups in that molecule}} \text{ g / mol}$$

Therefore, the EEW is one-half the average molecular weight of a diepoxy resin having linear chain with no branching and two epoxy groups at both ends and one-third the average molecular weight of a tri-epoxy resin that is epoxy resin with three end epoxide groups. Epoxide content is also expressed as the epoxide value (number, which is the number of epoxide equivalents in 100 grams of resin (eq/100g). The determination of the epoxide value of an epoxy resin or epoxy compound by chemical analysis (titration) method, commonly involves the use of hydrochloric acid – acetone mixture (2.8ml of 36.5% HCl in 140 ml of Acetone), the method is referred as hydrochloric acid-acetone-standing method. In this method the mixture of epoxy resin or epoxy compound with hydrochloric acid and acetone solution are left standing for at least 30 min. During this period the addition reaction of epoxide group with HCl takes place. The excess hydrochloric acid is then titrated with NaOH with respect to back titration. If X and Y are the volume of NaOH required for titration of unreacted HCl of Blank and Experimental titration respectively, then the epoxy equivalent weight (EEW) can be determined by using the following mathematical calculations,

1000 ml 1N HCl = 1000 ml 1N NaOH = 1 Gram equivalent of epoxy = Mol. wt. of polymer

(X – Y) ml of N_{NaOH} N NaOH = How much gram equivalent of epoxy

$$\text{Gram equivalent of epoxy} = \frac{N_{NaOH} * (X - Y) * 1}{1000 * 1}$$

Gram equivalent of epoxy per g of epoxy resin

$$= \frac{N_{NaOH} * (X - Y)}{1000 * \text{Weight of epoxy resin taken}}$$

Since epoxide equivalent weight (EEW) is the weight in grams of an epoxy resin containing one mole equivalent of epoxide (g/mol) hence epoxide equivalent weight (EEW) can be obtained by taking the reciprocal of above equation.

Epoxy Equivalent weight (EEW)

$$= \frac{1000 * \text{Weight of epoxy resin taken}}{N_{NaOH} * (X - Y)} = \text{g/mol}$$

$$\text{Epoxy Equivalent weight (EEW)} = \frac{1000 * W}{N_{NaOH} * (X - Y)} \ \text{g/mol}$$

Where W = Weight of epoxy resin taken

N_{NaOH} = Normality of standardized Sodium hydroxide solution

X = Volume of NaOH solution required for Blank titration

Y = Volume of NaOH solution required for Experimental titration

PROCEDURE

Standardized 0.1N NaOH solution by titrating it against standard oxalic acid solution and find out the accurate concentration of NaOH solution (N_{NaOH}). Weigh accurately 0.5 g of epoxy resin sample and transfer in 250 ml conical or Round Bottom flask. Add 25ml of Acetone – Hydrogen chloride solution mixture (2.8ml of 36.5% HCl in 140 ml of Acetone). Stoppard the flask with ground glass stopper. Shake the flask to dissolve the epoxy resin completely and to get clear solution. Remove the stopper, clean it with little quantity of acetone and transfer the washing to the flask. Add few drops of phenolphthalein indicator and titrate the solution against previously standardized 0.1N sodium hydroxide solution (EXPERIMENTAL). The end point is detected by the colour change from colourless to pink. Repeat the same procedure without taking the polymer sample (BLANK).

Note: The amount of specimen weight to be taken for the experiment is dependent on the expected epoxide equivalent weight (EEW) as follows:

EEW Approximate	Specimen Size g
170–375	0.4
375–600	0.6
600–1000	0.8
1000–1500	1.3
1500–2000	1.8
2000–2500	2.3
2500–5000	2.8

OBSERVATIONS

1. Weight of empty weighing bottle = W_1 g

2. Weight of weighing bottle with epoxy resin sample = W_2 g
3. Weight of epoxy resin sample taken = W = $(W_2 - W_1)$ g
4. Normality of standardized NaOH solution used = (N_{NaOH})
5. Volume of NaOH solution required for Experimental titration = Y ml
6. Volume of NaOH solution required for Blank titration = X ml

CALCULATIONS

Determine the Epoxy Equivalent weight (EEW) by using the following equation,

$$\text{Epoxy Equivalent weight (EEW)} = \frac{1000 * W}{N_{NaOH} * (X - Y)} \text{ g/mol}$$

RESULT

Epoxy Equivalent weight (EEW) of given epoxy resin is found to be = …

❑❑❑

EXPERIMENT NO. 14

AIM

Determination of epoxy equivalent weight (EEW) of an epoxy resin using in-situ hydrobromic acid.

APPARATUS

Electronics balance, Conical flask, Stopper, Burette, Pipette etc.

CHEMICALS

Given Epoxy resin, 60% Perchloric acid, Acetic acid, Methylene chloride, Potassium acid phthalate ($KHC_8H_4O_4$) primary standard grade. 20 % Tetra ethyl ammonium Bromide, acetic anhydride, Crystal violet indicator.

THEORY

The epoxy equivalent weight is significant since the physical properties of the resins such as flexibility, brittleness, permeability and adhesion are directly related to it. Kindly refer the theory given in determination of Epoxy Equivalent Weight (EEW) of an Epoxy resin in previous experiment.

In this method the resin is dissolved in a suitable solvent and the resulting solution is then titrated with hydrogen bromide either directly or in-situ. The direct titration method involves the titration of standard solution of hydrogen bromide in glacial acetic acid with epoxy groups to form bromohydrins .Where as in-situ method involves the titration of epoxy groups with standard perchloric acid in the presence of an excess of tetraethyl ammonium bromide. Hydrogen bromide generated in situ by the addition of perchloric acid to the quaternary ammonium halide rapidly opens the oxirane ring. Therefore, the quantity of hydrobromic acid consumed is a measure of the epoxy content of the resin.

The epoxy equivalent weight (EEW) using in-situ Hydrobromic acid can be determine by using the following mathematical calculations,

1000 ml 1N $HClO_4$ = 1 Gram equivalent of epoxy = Mol. wt. of polymer

V ml of N_{HClO4} N $HClO_4$ = How much gram equivalent of epoxy

$$\text{Gram equivalent of epoxy} = \frac{N_{HClO_4} * V * 1}{1000 * 1}$$

Gram equivalent of epoxy per g of epoxy resin

$$= \frac{N_{HClO_4} * V * 1}{1000 * \text{Weight of epoxy resin taken}}$$

Since epoxide equivalent weight (EEW) is the weight in grams of an epoxy resin containing one mole equivalent of epoxide (g/mol) hence epoxide equivalent weight (EEW) can be obtained by taking the reciprocal of above equation.

Epoxy Equivalent weight (EEW)

$$= \frac{1000 * \text{Weight of epoxy resin taken}}{N_{HClO_4} * V} = \text{g/mol}$$

$$\text{Epoxy Equivalent weight (EEW)} = \frac{1000 * W}{N_{HClO_4} * V} = \text{g/mol}$$

Where W = Weight of epoxy resin taken

N_{HCLO4} = Normality of standardized Perchloric acid solution

V = Volume of Perchloric acid solution required for titration

SOLUTION PREPARATION

1. *0.1 N Perchloric Acid solution in Glacial Acetic Acid:*

   Prepare the solution in the following manner and sequence in order to avoid an excessive rise in temperature.

   Place approximately 250 ml of glacial acetic acid into a 1000 ml volumetric flask. Add 13 ml of 60 % perchloric acid and mix well. Add 50 ml of acetic anhydride, dilute to the mark with glacial acetic acid and mix thoroughly. Allow to stand at least 8 hrs for completion of reaction between the acetic anhydride and water.

2. *20% Tetraethylammonium Bromide Solution in Glacial Acetic Acid:*

   Dissolve 100 g of tetraethylammonium bromide in 400 ml of glacial acetic acid with agitation at room temperature.

3. *Crystal Violet Indicator Solution*

   Prepare 0.1 % solution of crystal violet indicator in glacial acetic acid.

PROCEDURE

1. **Standardization of 0.1 *N* Perchloric acid solution ($HClO_4$)**

   Standardized the 0.1 N Perchloric acid solution by titrating it against Potassium acid phthalate as given below,

   Take accurately weighted 0.4 g of potassium acid phthalate in a beaker. Dissolve it in 50 ml of glacial acetic acid by slightly warming the solution if necessary and cool to room temperature. Add 6 to 8 drops of crystal violet indicator solution. Titrate with 0.1N perchloric acid solution to the end point, which is a sharp change in color from blue to green, stabilize for at least 2 min. Keep on stirring the solution using the magnetic stirrer during the course of titration.

   Calculate and record the normality of perchloric acid solution by using the following equation,

   Normality of Perchloric acid solution

$$(N_{HClO_4}) = \frac{1000 * W}{204.2 * V}$$

Where N = Normality of perchloric acid reagent,

W = Potassium acid phthalate used, g, and

V = Volume of perchloric acid reagent required to titrate (Burette reading)

2. **Procedure for determining Epoxy equivalent weight:**

   Take accurately weighed 0.5g of sample into a conical flask. (The weight of sample required is dependent on the epoxide equivalent weight (EEW) of the resin under study as given in previous experiment.) Add 10 to 15 ml of methylene chloride to the sample. Add 10 ml of tetra ethyl ammonium bromide reagent and 6 to 8 drops of crystal violet indicator solution. Titrate with standardized 0.1N perchloric acid solution to a sharp blue to green end point which is stable for at least 30 seconds. Record the burette reading of volume of perchloric acid solution required for the titration.

CALCULATION

Calculate the epoxy equivalent weight of epoxy resins (100% Solid) by using the following equation:

$$\text{Epoxy Equivalent weight (EEW)} = \frac{1000 * W}{N_{HClO_4} * V}$$

Where W = Weight of epoxy resin taken

N_{HCLO4} = Normality of standardized Perchloric acid solution

V = Volume of Perchloric acid solution required for titration

However the epoxy equivalent weight of epoxy resins (semi-solid) can be calculated by using the following equation,

Epoxy Equivalent weight (EEW)

$$= \frac{1000 * W}{N_{HClO_4} * V} * \frac{\text{Solid percentage of epoxy resin}}{100}$$

RESULT

Epoxy Equivalent weight (EEW) of given epoxy resin is found to be = ...

❑❑❑

EXPERIMENT NO. 15

AIM

Determination of epoxy equivalent weight (EEW) of an epoxy resin using hydrobromic acid.

APPARATUS

Electronics balance, Conical flask, Stopper, Burette, Pipette etc.

CHEMICALS

Given Epoxy resin, Acetic acid, Potassium Acid Phthalate ($KHC_8H_4O_4$) primary standard grade, Methyl isobutyl ketone, methyl violet indicator.

THEORY

The epoxy equivalent weight (EEW) by the direct use of Hydrobromic acid can be determine by using the following mathematical calculations,

1000 ml 1N HBr = 1 Gram equivalent of epoxy = Mol. wt. of polymer

V ml of N_{HBr} N HBr = How much gram equivalent of epoxy

$$\text{Gram equivalent of epoxy} = \frac{N_{HBr} * V * 1}{1000 * 1}$$

Gram equivalent of epoxy per g of epoxy resin

$$= \frac{N_{HBr} * V * 1}{1000 * \text{Weight of epoxy resin taken}}$$

Since epoxide equivalent weight (EEW) is the weight in grams of an epoxy resin containing one mole equivalent of epoxide (g/mol) hence epoxide equivalent weight (EEW) can be obtained by taking the reciprocal of above equation.

Epoxy Equivalent weight (EEW)

$$= \frac{1000 * \text{Weight of epoxy resin taken}}{N_{HBr} * V} \text{ g/mol}$$

$$\text{Epoxy Equivalent weight (EEW)} = \frac{1000 * W}{N_{HBr} * V} \text{ g/mol}$$

Where W = Weight of epoxy resin taken

N_{HBr} = Normality of standardized hydrobromic acid solution

V = Volume of Hydrobromic acid solution required for titration

SOLUTION PREPARATION

1. *0.1 N Hydrobromic Acid solution in Glacial Acetic Acid:* Prepare 0.1 N HBr in glacial acetic acid by adding 8 g of HBr to 500 ml of glacial acetic acid.
2. *Methyl Violet Indicator Solution* Prepare 0.1 % solution of methyl violet indicator in glacial acetic acid.

PROCEDURE

1. **Standardization of 0.1 *N* Hydrobromic acid (HBr):**

   Standardized the 0.1 N hydrobromic acid solution by titrating it against Potassium acid phthalate as given below.

   Take accurately weighted 0.4 g of potassium acid phthalate in a beaker. Dissolve it in 50 ml of glacial acetic acid by slightly warming the solution if necessary and cool to room temperature. Add 6 to 8 drops of methyl violet indicator solution. Titrate with 0.1N hydrobromic acid solution to the end point, which is a sharp colour change from purple to light blue and finally to green. Keep on stirring the solution using the magnetic stirrer during the course of titration.

   Calculate and record the normality of Hydrobromic acid solution by using the following equation,

   Normality of Hydrobromic acid solution

$$(N_{HBr}) = \frac{1000 * W}{204.2 * V}$$

Where N = Normality of *Hydrobromic* acid solution

W = Potassium acid phthalate used in g

V = Volume of *Hydrobromic* acid solution required for titration (Burette reading)

2. **Procedure for determining Epoxy equivalent weight:**

   Take accurately weighed 0.5g of sample into a conical flask. (The weight of sample required is dependent on the epoxide equivalent weight (EEW) of the resin under study as given in previous experiment.) Add 10 to 15 ml of methyl isobutyl ketone to dissolve the sample. Add 10 ml of glacial acetic acid just before the titration and 6 to 8 drops of methyl violet indicator solution. Titrate with standardized 0.1N hydrobromic acid solution to a sharp colour change from purple to light blue to finally green. Record the burette reading of volume of hydrobromic acid solution required for the titration.

CALCULATION

Calculate the epoxy equivalent weight of epoxy resins (100% Solid) by using the following equation:

$$\text{Epoxy Equivalent weight (EEW)} = \frac{1000 * W}{N_{HBr} * V}$$

Where W = Weight of epoxy resin taken

N_{HBr} = Normality of standardized hydrobromic acid solution

V = Volume of hydrobromic acid solution required for titration

However the epoxy equivalent weight of epoxy resins (semi-solid) can be calculate by using the following equation,

Epoxy Equivalent weight (EEW)

$$= \frac{1000 * W}{N_{HBr} * V} * \frac{\text{Solid percentage of epoxy resin}}{100}$$

Note: Where Solid percentage of epoxy resin is provided by the manufacturer.

RESULT

Epoxy Equivalent weight (EEW) of given epoxy resin is found to be =...

❑❑❑

EXPERIMENT NO. 16

AIM

Determination of amine value of polyamide resin using in-situ hydrobromic acid.

APPARATUS

Electronics balance, conical flask, Stopper, Burette, Pipette etc.

CHEMICALS

Given polyamide resin, 60% Perchloric acid, Acetic acid, Methylene chloride, Potassium Acid Phthalate ($KHC_8H_4O_4$) primary standard grade. 20 % Tetra ethyl ammonium Bromide, acetic anhydride, Crystal violet indicator.

THEORY

Polyamides are the most commonly used epoxy curing agent. Curing is a term in polymer chemistry and process engineering that refers to the toughening or hardening of a polymer material by cross-linking of polymer chains, brought about by electron beams, heat or chemical additives. The chemical additives includes the variety of compounds like anhydrides, phenolic resins and Lewis acids, however selecting the polymides as a curing for an epoxy resins are found to be the best due to low cost, easy processing requirements and performance requirements. The sensitivity of this method decreases as the molecular weight increases. Thus this technique is limited to the determination of polymers with a molecular weight of less than about 20,000.

The amine value gives the active hydrogen (NH) content of an amine *hardener*. It is expressed in terms of the weight of KOH in milligrams that require to neutralise the NH in 1 gram of polyamide resin. The *equivalent weight* of KOH is 56.1, hence

$$\text{Amine value} = \frac{N_{HClO_4} * V_{HClO_4} * 56.1}{W}$$

Where W = Weight of polyamide resin taken

N_{HCLO_4} = Normality of standardized Perchloric acid solution

V_{HCLO_4} = Volume of Perchloric acid solution required for titration

However the amine value of polyamide resins (semi-solid) can be calculate by using the following equation,

Amine value =

$$\frac{N_{HClO_4} * V_{HClO_4} * 56.1}{W} * \frac{100}{\text{Solid percentage of polymide resin}}$$

SOLUTION PREPARATION

1. *0.1 N Perchloric Acid solution in Glacial Acetic Acid:*

   Prepare the solution in the following manner and sequence in order to avoid an excessive rise in temperature.

   Place approximately 250 ml of glacial acetic acid into a 1000 ml volumetric flask. Add 13 ml of 60 % perchloric acid and mix well. Add 50 ml of acetic anhydride, dilute to the mark with glacial acetic acid and mix thoroughly. Allow to stand at least 8 hrs for completion of reaction between the acetic anhydride and water.

2. *20% Tetraethylammonium Bromide Solution in Glacial Acetic Acid:*

   Dissolve 100 g of tetraethylammonium bromide in 500 ml of glacial acetic acid with agitation at room temperature.

3. *Crystal Violet Indicator Solution*

   Prepare 0.1 % solution of crystal violet indicator in glacial acetic acid

PROCEDURE

1. **Standardization of 0.1 *N* Perchloric acid solution ($HClO_4$)**

   Standardized the 0.1 N Perchloric acid solution by titrating it against Potassium acid phthalate as given below,

Take accurately weighted 0.4 g of potassium acid phthalate in a beaker. Dissolve it in 50 ml of glacial acetic acid by slightly warming the solution if necessary and cool to room temperature. Add 6 to 8 drops of crystal violet indicator solution. Titrate with 0.1N perchloric acid solution to the end point, which is a sharp change in color from blue to green, stabilize for at least 2 min. Keep on stirring the solution using the magnetic stirrer during the course of titration.

Calculate and record the normality of perchloric acid solution by using the following equation,

Normality of Perchloric acid solution

$$(N_{HClO_4}) = \frac{1000 * W}{204.2 * V}$$

Where N = Normality of perchloric acid reagent,

W = Potassium acid phthalate used in g

V = Volume of perchloric acid reagent required to titrate (Burette reading)

2. **Procedure for determining amine value of polyamide resin:**

   Take accurately weighed 0.5g of sample into a conical flask. Add 10 to 15 ml of methylene chloride to the sample. Add 10 ml of tetra ethyl ammonium bromide reagent and 6 to 8 drops of crystal violet indicator solution. Titrate with standardized 0.1N perchloric acid solution to a sharp blue to green end point which is stable for at least 30 seconds. Record the burette reading of volume of perchloric acid solution required for the titration.

CALCULATION

Calculate the amine value if polyamide resins (100% Solid) by using the following equation:

$$\text{Amine value} = \frac{N_{HClO_4} * V_{HClO_4} * 56.1}{W}$$

Where W = Weight of polyamide resin taken

N_{HCLO4} = Normality of standardized Perchloric acid solution

V_{HCLO4} = Volume of Perchloric acid solution required for titration

However the amine value of polyamide resins (semi-solid) can be calculated by using the following equation,

Amine value =

$$\frac{N_{HC1O_4} * V_{HC1O_4} * 56.1}{W} * \frac{100}{\text{Solid percentage of polymide resin}}$$

RESULT

Amine value of given polyamide resin is found to be = ...

❑❑❑

EXPERIMENT NO. 17

AIM

Determination of amine value of polyamide resin using hydrobromic acid.

APPARATUS

Electronics balance, Conical flask, Stopper, Burette, Pipette etc.

CHEMICALS

Given polyamide resin, Acetic acid, Potassium Acid Phthalate ($KHC_8H_4O_4$) primary standard grade, Methyl isobutyl ketone, Methyl violet indicator.

THEORY

The amine value gives the active hydrogen (NH) content of an amine *hardener*. It is expressed in terms of the weight of KOH in milligrams that require to neutralise the NH in 1 gram of polyamide resin. The *equivalent weight* of KOH is 56.1, hence

$$\text{Amine value} = \frac{N_{HBr} * V_{HBr} * 56.1}{W}$$

Where W = Weight of polyamide resin taken

N_{HBr} = Normality of standardized hydrobromic acid solution

V_{HBr} = Volume of hydrobromic acid solution required for titration

However the amine value of polyamide resins (semi-solid) can be calculated by using the following equation,

$$\text{Amine value} = \frac{N_{HBr} * V_{HBr} * 56.1}{W} * \frac{100}{\text{Solid percentage of polymide resin}}$$

SOLUTION PREPARATION

1. *0.1 N Hydrobromic Acid solution in Glacial Acetic Acid:*

   Prepare 0.1 N HBr in glacial acetic acid by adding 8 g of HBr to 500 ml of glacial acetic acid.

2. *Methyl Violet Indicator Solution*

   Prepare 0.1 % solution of methyl violet indicator in glacial acetic acid

PROCEDURE

1. **Standardization of 0.1 *N* Hydrobromic acid (HBr):**

   Standardized the 0.1 *N* hydrobromic acid solution by titrating it against Potassium acid phthalate as given below,

   Take accurately weighted 0.4 g of potassium acid phthalate in a beaker. Dissolve it in 50 ml of glacial acetic acid by slightly warming the solution if necessary and cool to room temperature. Add 6 to 8 drops of methyl violet indicator solution. Titrate with 0.1N hydrobromic acid solution to the end point, which is a sharp colour change from purple to light blue and finally to green. Keep on stirring the solution using the magnetic stirrer during the course of titration.

   Calculate and record the normality of Hydrobromic acid solution by using the following equation,
   Normality of Hydrobromic acid solution

$$(N_{HBr}) = \frac{1000 * W}{204.2 * V}$$

   Where N = Normality of *Hydrobromic* acid solution
   W = Potassium acid phthalate used in g
   V = Volume of *Hydrobromic* acid solution required for titration (Burette reading)

2. **Procedure for determining amine value:**

   Take accurately weighed 0.5g of sample into a conical flask. Add 10 to 15 ml of methyl isobutyl ketone to dissolve the sample. Add 10 ml of glacial acetic acid just before the titration and 6 to 8 drops of methyl violet indicator solution. Titrate with standardized 0.1N hydrobromic acid solution. The endpoint for the amine value is not the same as for the epoxide procedure. In case of amine value, the amine value solution turns milky, then finally to green. Record the burette reading of volume of hydrobromic acid solution required for the titration.

CALCULATION

Calculate the amine value of polyamide resins (100% Solid) by using the following equation:

$$\text{Amine value} = \frac{N_{HBr} * V_{HBr} * 56.1}{W}$$

Where W = Weight of polyamide resin taken

N_{HBr} = Normality of standardized hydrobromic acid solution

V_{HBr} = Volume of hydrobromic acid solution required for titration

However the amine value of polyamide resins (semi-solid) can be calculated by using the following equation,

Amine value =

$$= \frac{N_{HBr} * V_{HBr} * 56.1}{W} * \frac{100}{\text{Solid percentage of polymide resin}}$$

Note: Where Solid percentage of polyamide resin is provided by the manufacturer.

RESULT

The amine value of given polyamide resin is found to be = …

❑❑❑

EXPERIMENT NO. 18

AIM

Determination of weight percent hydrolysable chlorine content of an epoxy resin.

APPARATUS

Electronics balance, Iodine flask, Water condenser, Stopper, Burette, Pipette etc.

CHEMICALS

Given Epoxy resin, 0.5N Methanolic Potassium hydroxide solution, 0.5N Hydrochloric acid, Toluene, Ethanol, Phenolphthalein indicator, Methyl ethyl ketone.

THEORY

The hydrolyzable chloride content of epoxy resins is an important variable in determining their reactivity and the properties of coatings made from them. Epoxy resins synthesized by epichlorohydrine method contains some active hydrolysable chlorine as 1, 2 chlorohydrine due to the incomplete dehydrohalogenation. The present method may be used to determine the hydrolyzable chloride content of manufactured epoxy resins and confirm the stated hydrolyzable chloride content of purchased epoxy resins. Determination of hydrolysable chlorine content in an Epoxy resin involves the saponification of resin under study with potassium hydroxide followed by direct titration of excess or unreacted potassium hydroxide against hydrochloric acid. The amount of potassium hydroxide consumed in the hydrolysis is a measure of the hydrolyzable chloride content of the resin. Thus hydrolysable chlorine content of an Epoxy resin then can be determined with respect to back titration. If X and Y are the volume of HCl required for titration of excess KOH of Blank and Experimental titration respectively, then the hydrolysable chlorine content can be determined by using the following mathematical calculations,

$$1000 \text{ ml of 1N HCL} = 35.5\text{g of chlorine}$$

$$(X - Y) \text{ ml of } N_{HCl} = \text{How much g of chlorine}$$

$$\text{Chlorine content in g} = \frac{(X - Y) * N_{HCl} * 35.5}{1000}$$

This is the chlorine content in g corresponding to the weight of sample taken (W).

Therefore the weight percentage Hydrolysable chlorine content of the epoxy resin is given by the equation,

Weight Percentage Hydrolysable chlorine content =

$$\frac{(X - Y) * N_{HCl} * 35.5 * 100}{1000 * \text{Weight of epoxy resin taken (W)}} \%$$

Weight Percent Hydrolysable chlorine content =

$$\frac{(X - Y) * N_{HCl} * 3.55}{\text{Weight of epoxy resin taken (W)}} \%$$

Where W = Weight of epoxy resin taken

N_{HCl} = Normality of standardized Hydrochloric acid solution

X = Volume of HCl solution required for Blank titration

Y = Volume of HCl solution required for Experimental titration

PROCEDURE

Standardized 0.5N KOH methanolic solution by titrating it against standard oxalic acid solution and find out the accurate concentration of KOH solution. Then standardized the 0.5N HCl solution by titrating it against above standardized KOH solution. Weigh and transfer accurately 2 -3g of epoxy resin sample into a 250mL glass-stoppered iodine flask. By means of a pipette, transfer 50ml of 0.5 N alcoholic KOH solution into the flask and add 15 ml of toluene. Stopper the flask and shake well to dissolve the sample completely. Attach flask to a reflux condensers and reflux gently for 15 minutes. At the end of the reflux period, cool the assembly to room temperature. Rinse down the condenser with 20 ml of toluene into the flask. Then remove the condensers from the flasks. Add few drops of phenolphthalein indicator solution to flask, it turns pink. Titrate the flask content with 0.5 N HCl. The end point is detected when the colour of solution changes from pink to colorless (EXPERIMENTAL). Repeat the same procedure without taking the polymer sample (BLANK).

Note: Add 100 mL of Methyl ethyl ketone (MEK) to the specimen if required to ensure a homogeneous solution during titration of the excess KOH solution.

OBSERVATIONS

1. Weight of epoxy resin sample taken = ... Wg
2. Normality of standardized KOH solution used= (N_{KOH})
3. Normality of standardized HCl solution used= (N_{HCl})
4. Volume of HCl solution required for Experimental titration = Y ml
5. Volume of HCl solution required for Blank titration = X ml

CALCULATIONS

Determine the Percentage Hydrolysable chlorine content by using the following equation,

Weight Percent Hydrolysable chlorine content =

$$\frac{(X - Y) * N_{HCl} * 3.55}{\text{Weight of epoxy resin taken (W)}} \%$$

RESULT

Weight percentage Hydrolysable chlorine content of given epoxy resin is found to be = ...

❑❑❑

EXPERIMENT NO. 19

AIM

Estimation of the vinyl monomer concentration by the bromination method.

APPARATUS

250ml Conical flask (Iodine flask), Pipette, Burette, etc.

CHEMICALS

Monomer solution (vinyl monomer), Brominating solution, 20% KI solution, 0.15M Sodium thiosulphate solution, 4% Sulphuric acid solution, Starch Indicator.

PROCEDURE: (A) Blank Titration

Standardization of Brominating Solution

Pipette out 25ml of brominating solution in an iodine/ conical flask. To it add 10ml of 4% Sulphuric acid solution and keep the flask as such for 15 minutes. Then add 25ml of 20% potassium iodide solution. Titrate the complete solution against standard 0.15M Sodium thiosulphate solution using starch solution as an indicator. The addition of starch indicator to the titration flask should be made after the solution became pale yellow on titration with sodium thiosulphate. After addition of starch the mixture turns violet. Further titration to a colour change from violet blue to colourless indicates the end point of the titration (V_1).

(B) Experimental Titration

Estimation of the vinyl monomer

Dilute the given amount of monomer solution by distilled water to 100ml in a volumetric flask. Pipette out 10ml of this solution in an iodine/conical flask and add 25ml of brominating solution and then 10ml of 4% Sulphuric acid solution to the flask. Keep the flask as such for 15 minutes. Then add 25ml of 20% potassium iodide solution. Titrate the complete solution against standard 0.15M Sodium thiosulphate solution using starch solution as an indicator as explain above (V_2).

OBSERVATIONS

1. Volume of sodium thiosulphate required for the Blank Titration (A) = V_1 ml
2. Volume of sodium thiosulphate required for the Experimental Titration (B) = V_2 ml

CALCULATIONS

Concentration of monomer =

$$\frac{(V_1 - V_2) * \text{ Molarity of } Na_2S_2O_3}{\text{Volume of monomer solution} * 2}$$

RESULT

The concentration of monomer = ... Moles/lit.

❑❑❑

EXPERIMENT NO. 20

AIM

Determination of Molecular weight (Mn) of polyethylene glycol by end group analysis.

APPARATUS

250ml Conical or Round Bottom flask with joint, Reflux Condenser, Water Bath, Pipette, Burette, Measuring Cylinder.

CHEMICALS

Polyethylene Glycol, Acetylating mixture (*i.e.* 1:1 ratio of Acetic acid and Acetic anhydride), Methanol, 1N Potassium hydroxide, Phenolphthalein Indicator.

THEORY

End Group Analysis is a chemical method used for calculating the number average molecular weight of polymer containing reactive functional groups at it's one or both the ends. In Polyethylene Glycol hydroxyl group is present at its one end. The number of hydroxyl groups gives the information about the Number average molecular weight of Polyethylene Glycol polymer.

The determination is carried out by acetylating polymer with acetic anhydride in acetic acid and resulting mixture is then titrated with standardized alkali (KOH) solution.

PROCEDURE

Weigh accurately 0.25 g of polymer sample and transfer in 250 ml conical or Round Bottom flask. Add 5ml of freshly prepared acetylating mixture and put porcelain pieces in the flask. Reflux the mixture in water bath using water condenser. Continue the refluxing for one hour. The flask allows to cool to room temperature and then add 5ml of water from the side of the flask and again reflux for 5 minutes. Cool it again and add 25ml of Methanol from the side of the flask. Shake well the mixture and titrate the complete mixture against 1N KOH solution using Phenolphthalein indicator (EXPERIMENTAL). Repeat the same procedure without taking the polymer sample (BLANK). The end point is detected by the colour change from colourless to pink.

REACTIONS

Polymer-OH + Ac_2O → Polymer-OAC + ACOH

$Ac_2O + H_2O \rightarrow$ 2ACOH

ACOH + KOH → ACOK + H_2O

OBSERVATIONS

1. Standardized 1N KOH solution by titrating it against standard oxalic acid solution and find out the accurate concentration of KOH solution used (N_{KOH}).
2. Burette reading for Experimental titration:- Y ml of KOH
3. Burette reading for Blank titration:- X ml of KOH

CALCULATIONS

Number of Hydroxyl groups in mg of KOH (A) =

$$\frac{56.1 \times \text{N of KOH} \times (X - Y)}{\text{Weight of polymer sample taken}}$$

And

Number average molecular wt. (M_n) =

$$\frac{56100 * 2}{\text{Number of Hydroxyl groups (A)}}$$

RESULT

The Number of Hydroxyl groups in given polymer sample (A) = ...

The Number average molecular weight of given polymer sample (Mn) = ...

❑❑❑

EXPERIMENT NO. 21

AIM

Determination of Molecular weight (Mn) of Polyvinyl alcohol by hydroxyl end Group Analysis.

APPARATUS

250ml Conical or Round Bottom flask with joint, Reflux Condenser, Water Bath, Pipette, Burette, Measuring Cylinder.

CHEMICALS

Polyvinyl alcohol, Acetic anhydride, Pyridine, 1N Potassium hydroxide, Phenolphthalein Indicator.

THEORY

End Group Analysis is a chemical method used for calculating the number average molecular weight of polymer containing reactive functional groups at it's one or both the ends. In Polyvinyl alcohol hydroxyl group is present at its one end. The number of hydroxyl groups gives the information about the Number average molecular weight of polyvinyl alcohol polymer.

The determination is carried out by acetylating polymer with acetic anhydride in pyridine and resulting mixture is then titrated with standardized alkali (KOH) solution.

PROCEDURE

Weigh accurately 0.25 g of polymer sample and transfer in 250 ml conical or Round Bottom flask. Add 5ml of acetic anhydride and 10ml of pyridine. Put porcelain pieces in the flask and reflux the mixture in water bath using water condenser. Continue the refluxing for half hour. The flask allow to cool to room temperature and then add 70ml of water from the side of the flask. Shake well the mixture and titrate the complete mixture against 1N KOH solution using Phenolphthalein indicator (EXPERIMENTAL). Repeat the same procedure without taking the polymer sample (BLANK). The end point is detected by the color change from colorless to pink.

REACTIONS

$$\text{Polymer-OH} + Ac_2O \rightarrow \text{Polymer-OAC} + \text{ACOH}$$

$$Ac_2O + H_2O \rightarrow 2\text{ACOH}$$

$$\text{ACOH} + \text{KOH} \rightarrow \text{ACOK} + H_2O$$

OBSERVATIONS

1. Standardized 1N KOH solution by titrating it against standard oxalic acid solution and find out the accurate concentration of KOH solution used (N_{KOH}).
2. Burette reading for Experimental titration:- Y ml of KOH
3. Burette reading for Blank titration:- X ml of KOH

CALCULATIONS

Number of Hydroxyl groups in mg of KOH (A) =

$$\frac{56.1 * \text{N of KOH} * (X - Y)}{\text{Weight of polymer sample taken}}$$

And

Number average Molecular wt. (M_n) =

$$\frac{56100 * 2}{\text{Number of Hydroxyl groups (A)}}$$

RESULT

The Number of Hydroxyl groups in given polymer sample (A) = ...

The Number average molecular weight of given polymer sample (Mn) = ...

❑❑❑

EXPERIMENT NO. 22

AIM

Determination of Number average Molecular weight (Mn) of Nylon-6 by End Group Analysis.

APPARATUS

250ml beaker, Pipette, Microburette, Measuring Cylinder, Conductivity meter.

CHEMICALS

Nylon-6, 0.1N HCL, 0.1N NaOH, 95% Ethanol, Benzyl alcohol, Al-foil, Phenolphthalein Indicator.

THEORY

End Group Analysis is a chemical method used for calculating the number average molecular weight of polymer containing reactive functional groups at it's one or both the ends. In Nylon-6 polymer amine and carboxylic functional groups are present at its two ends.

The amine end group can be determine by conductometric titration with standardized hydrochloric acid solution where as the carboxyl end group can be determined by titration with standard alkali solution.

PROCEDURE

PART-A: Determination of Amine end group by Conductometrically.

Weigh accurately 2 g of polymer sample and dissolve in 50ml of phenol in a beaker. Add 25ml of 95% ethanol and 25ml of water. Titrate the mixture conductometrically with 0.1N HCL using microburett by successive addition of 0.1ml HCL solution. Continue titration after end point with excess addition of HCL.

PART-B: Determination of Carboxyl end group

Take accurately weighted 2 g of polymer sample and 50ml of benzylalcohol in a beaker. Covered the beaker with Al-foil to minimized vapour losses and oxidation of alcohol. Place the beaker in benzylalcohol bath maintained at 185-200°C temperature. Insert the stirrer through a small hole in Al-foil and stirred the mixture continuously for 45 mints to dissolve the sample completely. When the polymer get completely dissolved remove the Al-foil cover, add few drops of phenolphthalein indicator and then titrate the complete solution with 0.1N KOH solution (EXPERIMENTAL). Repeat the same procedure without taking the polymer sample (BLANK is run on Benzyl alcohol). The end point is detected by the colour change from colourless to pink.

OBSERVATIONS

PART-A: Determination of Amine end group by Conductometrically.

1. Weight of sample taken: W_1 gm
2. Normality of HCL Solution: N_{HCL}

PART-B: Determination of Carboxyl end group.

1. Weight of sample taken: W_2 gm
2. Standardized 0.1N KOH solution by titrating it against standard oxalic acid solution and find out the accurate concentration of KOH solution used: (N_{KOH}).
3. Burette reading for Experimental titration:- Y ml of KOH
4. Burette reading for Blank titration:- X ml of KOH

CALCULATIONS

PART-A: Determination of Amine end group by Conductometrically.

From graph the volume of HCL corresponding to the end point of titration: V ml

Amine End equivalence per 10^6 gm (AE)

$$AE = \frac{V * N \text{ of HCL} * 1000}{\text{Weight of polymer sample taken } (W_1)}$$

PART-B: Determination of Carboxyl end group.

Carboxyl End equivalence per 10^6 gm (CE)

$$CE = \frac{N \text{ of KOH} * (X - Y) * 1000}{\text{Weight of polymer sample taken } (W_2)}$$

Calculation of Number Average Molecular Weight (Mn) from Part A & Part B

$$Mn = \frac{10^6 \times 2}{\text{Amine End equivalence (AE)} + \text{Carboxyl End equivalence (CE)}}$$

RESULT

The Number average Molecular weight (Mn) of Nylon-6 by End Group Analysis = ...

❑❑❑

EXPERIMENT NO. 23

AIM

Determination of weight average molecular weight (Mw) of given polymer sample by viscosity method (Polyvinyl alcohol).

APPARATUS

Ubbelohde viscometer, stop watch, Beaker.

CHEMICALS

Polyvinyl alcohol, distilled water.

THEORY

The physical property viscosity of the polymer solution is useful to measure the molecular weight of polymer. The viscosity of the polymer solution is basically measure of the size or extension in space of polymer molecules. It is related to the molecular weight of linear polymers only. The relationship of average molecular weight of polymer with the intrinsic viscosity [η] of polymer solution is given by Mark Houwink's equation as

$$[\eta] = K\,M^{\alpha}$$

Where M is the average molecular weight of polymer.

[η] is the intrinsic viscosity of the polymer solution. It can be obtained from the plot of $\frac{\eta_{sp}}{C}$ along Y-axis against % concentration along X-axis and extrapolating the straight line to zero concentration i.e.

$$[\eta] = \lim_{c \to 0} \left(\frac{\eta_{sp}}{C}\right)$$

K & α are Mark Houwink's constants for given polymer and solvent pair at the temperature of viscosity measurement.

The values of K and α for Polyvinyl alcohol and water solvent are,

$$K = 4.53 \times 10^{-4}\ \text{ml/g and } \alpha = 0.64$$

For different pair of polymer and solvent system, the values of K & α are given in following table.

Polymer and solvent system	K X 10^{-4} ml/g	α
Polystyrene-Benzene	1.06	0.735
Polystyrene-Toluene	1.10	0.725
Polyvinyl acetate-Acetone	1.02	0.72
Polyvinyl acetate- Benzene	5.63	0.62
Polyvinyl acetate-Acetonitrile	4.15	0.62
Polyvinyl acetate-Methanol	3.14	0.60
Polymethylmethacrylate-Benzene	0.52	0.76
Polymethylmethacrylate- Acetone	0.77	0.70
Polymethylmethacrylate- Toluene	0.70	0.71
Polyvinyl Chloride-THF	8.33	0.83
Polyvinyl Chloride-Cyclohexanone	1.63	0.77
Polyacrylamide-Water	6.80	0.66
Nylon 6-M-cresol	32.00	0.62
Nylon 66-M-cresol	24	0.61
Cellulose acetate-Acetone	1.49	0.82
Polyisobutylene-Toluene	3.6	0.64

PROCEDURE

Prepare the 1% solution of the given polymer sample (Polyvinyl alcohol) in suitable solvent (water). Clean the Ubbelohde viscometer first with chromic acid and then with acetone. Clamp the viscometer properly in vertical position with the help of stand. Fill it with the solvent (water) and determine the flow time in between two marks above and below the upper bulb (**to**). Remove the water from the viscometer. Rinse & fill it with known minimum required exact volume of 1% polymer solution and determine the flow time. Dilute the polymer solution successively inside the viscometer to the concentrations 0.8, 0.6, 0.4 & 0.2% by adding calculated exact volume of solvent (water). Determine the flow time for each concentrations (***t***).

If the volume of solution inside the viscometer at any stage of dilution do not permit the solution to flow in between the two marks, remove the known sufficient volume of the solution from the side outlet of the viscometer and then dilute remaining known volume of the solution inside the viscometer by adding the calculated exact volume of the solvent to get the required concentration.

OBSERVATIONS

Room temperature

Flow time for solvent (Water) *i.e.* to = … sec

Sr. No.	Concentration g/ 100ml C (%)	Flow Time in sec.				Relative viscosity	Specific viscosity	
		t_1	t_2	t_3	Mean (t)	$\eta_r = \frac{t}{to}$	$\eta_{sp} = (\eta_r - 1)$	$\frac{\eta_{sp}}{c}$
1	1							
2	0.8							
3	0.6							
4	0.4							
5	0.2							

CALCULATIONS

1. Plot the graph between $\frac{\eta_{sp}}{c}$ along Y-axis and % concentration along X-axis. Determine intrinsic viscosity [η] of the polymer solution by extrapolating the straight line obtained to zero concentration.

2. Calculate average molecular weight of the polymer (M) by substituting the literature values of K, α and graphical value of intrinsic viscosity [η] using the following equation.

$$[\eta] = K\,M^{\alpha}$$

RESULT

The weight average molecular weight (Mw) of given polymer sample by viscosity method =

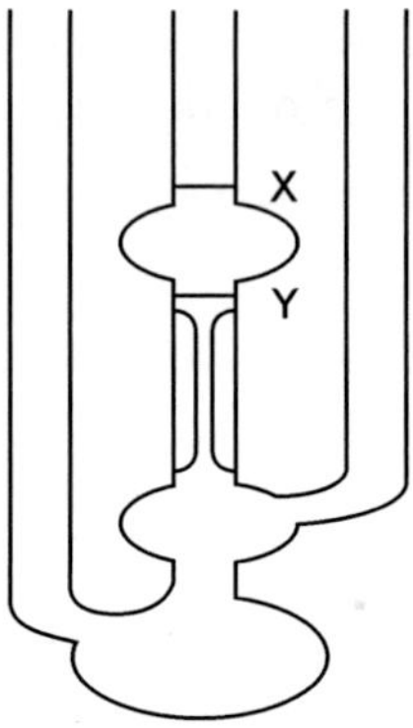

FIG. 5.0. Sketch of Ubbelohde viscometer

❑❑❑

EXPERIMENT NO. 24

AIM

Determination of Number Average Molecular Weight (Mn) of polystyrene by measurement of osmotic pressure.

APPARATUS

Membrane Osmometer, Volumetric flask, Pipette, Stop clock etc.

CHEMICALS

Polystyrene, Toluene.

THEORY

The two common osmometry techniques used for number average molecular weight determination of the polymers are mainly the membrane osmometry and vapour pressure osmometry. Membrane Osmometer technique is based on the phenomenon of osmosis. In this technique the polymer solution and pure solvent are separated by a membrane which is semi permeable only to the solvent and not to the polymer (solute). Because the chemical potential of the pure solvent is higher than that of the solvent in polymer solution, the solvent moves from the pure solvent side to the polymer solution side. This will be reflected as the rise of the solution level in the capillary attached to the solution compartment. The process will continue till the hydrostatic pressure developed in the capillary due to increase in the level of solution exactly balances the osmotic pressure. Once the osmotic equilibrium is achieved, the solvent stops entering the solution side. Under this situation the difference in the height of solution and the solvent levels in the two capillaries give the osmotic head (Δh) of the solution at that concentration.

The osmotic head (Δh) is related to the osmotic pressure (π) of the solution by the following equation,

$$\pi\ (g/cm^2) = \Delta h\ (cm) * d\ (g/cm^3)$$

Since 1 atmosphere = 1033 g/cm^2,

Hence osmotic pressure (π) in atmosphere is given by the following equation,

$$\pi \text{ (atm.)} = \frac{\Delta h * d}{1033}$$

Where π = Osmotic pressure of polymer solution in atm.

d = Density of polymer solution in g/cm^3

Δh = Osmotic head of polymer solution in cm

Now the reduced osmotic pressure (π/c) is related to the number average molecular weight of the polymer by the following equation,

$$(\pi \mid c)_{c \to 0} = \frac{RT}{Mn}$$

Where $(\pi \mid c)_{C=0}$ = Limiting osmotic pressure of solution

C = Concentration of polymer solution in g/100 ml

R = Gas constant, 0.8205 Lit. atm $mol^{-1}K^{-1}$.

Mn = The number average molecular weight of polymer solution

T = Temperature of measurement in Kelvin

Thus the plot of reduced osmotic pressure (π/c in atm.) along Y – axis against the percentage concentration (g/100ml)

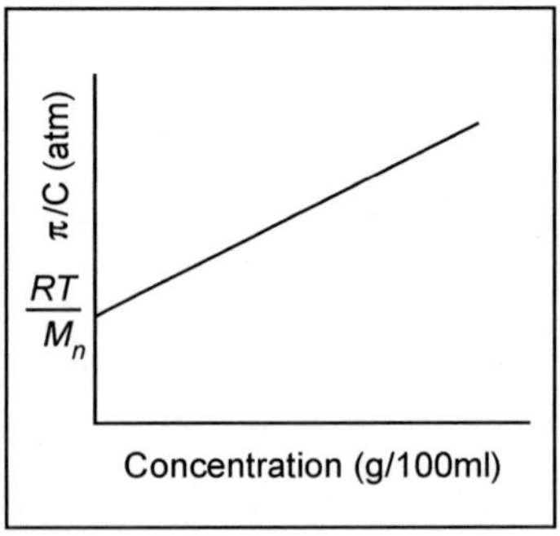

FIG. 5.1. Plot of reduced osmotic pressure (π/c in atm.) versus concentration (g/100ml). (Exp. No. 24)

along X-axis and Extrapolating the straight line obtained to zero concentration *i.e.* $(\pi/c)_{C=0}$ as shown in the figure 5.1 can be used to determine the Number average molecular weight of the polymer.

PROCEDURE

Weigh accurately 1.0 g of polystyrene into a 100 ml volumetric flask. Add about 50 ml toluene to the sample and shake the mixture until all polystyrene get dissolved and form a homogenous solution. Make up the volume of the solution with toluene to the 100 ml graduated mark. The concentration of this stock solution is 1% (*i.e.* 1 g/100 ml)

Now prepare the polystyrene solutions of different percentage concentrations (*i.e.* 0.8, 0.6, 0.4, and 0.2 %) by the method of dilution using stock solution. Determine the density of pure solvent and that of the each solution. Record the osmotic head (Δh cm) value using Osmometer instrument for different concentrations of polymer solution. Plot the graph of reduced osmotic pressure (π/c in atm.) along Y – axis against the percentage concentration (g/100ml) along X-axis. Extrapolate the straight line obtained to zero concentration and determine the Number average molecular weight by using the above equation. The density of the solutions can be taken as density of solvent since the solutions are very much dilute.

OBSERVATIONS

1. Room Temperature in Kelvin T =
2. Density of pure solvent (Toluene) =

OBSERVATION TABLE

Sr. No.	% Concentration (C)	Density (d g/ml)	Osmotic head (Δh cm)	Osmotic Pressure π (atm.) $= \frac{\Delta h * d}{1033}$	π/c
1	1				
2	0.8				
3	0.6				
4	0.4				
5	0.2				

CALCULATIONS

Plot the graph of reduced osmotic pressure (π/c in atm.) along Y – axis against the percentage concentration (g/100ml) along X-axis. Extrapolate the straight line obtained to zero concentration and determine the Number average molecular weight (Mn) by using the following equation.

$$(\pi \mid c)_{c \to 0} = \frac{RT}{Mn}$$

RESULT

The Number Average Molecular Weight (Mn) of polystyrene by measurement of osmotic pressure is found to be ...

❑❑❑

EXPERIMENT NO. 25

AIM

Determination of glass transition temperature of polymer by cooling curve method.

APPARATUS

Boiling test tube, Stop watch, Glass wool, Thermometer etc.

CHEMICALS

Polymer sample.

THEORY

In the study of polymers and their applications, it is important to understand the concept of the glass transition temperature (Tg). As the temperature of a polymer drops below Tg, it behaves in an increasingly brittle manner. As the temperature rises above the Tg, the polymer becomes more rubber-like. Thus, knowledge of Tg is essential in the selection of materials for various applications. Many physical properties change profoundly at the glass transition temperature, including the coefficient of thermal expansion, heat capacity, refractive index, mechanical and electrical properties which are important for the applications of materials in the field of chemistry, food, pharmacy and optics etc. Thus Glass transition is a "pseudo" second order phase transition. Numerous experimental techniques have been proposed for the determination of glass transition temperature.

In the present cooling curve method, the fall in temperature of the molten polymer as a function of time is used for the determination of glass transition temperature of polymer. As the transition of the polymer from molten state to the glassy state occurs over the certain range of temperature which include Tg. Therefore change in the slope of line due to glass transition which occurs during cooling gives approximate value of glass transition temperature as shown in the following graph Fig. 5.2.

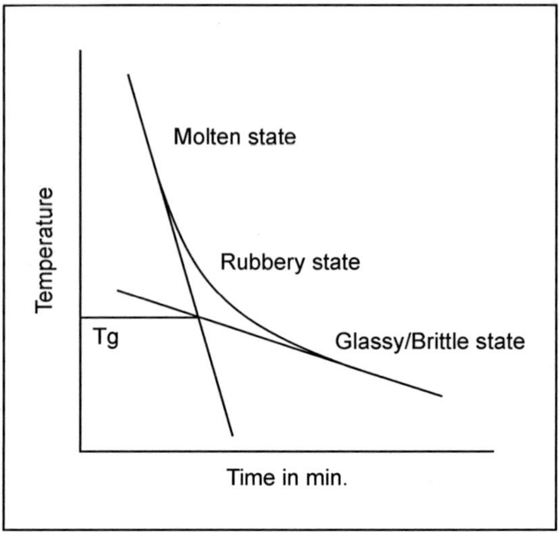

FIG. 5.2. Glass transition temperature by cooling curve.

PROCEDURE

Take about 5 g of powdered or small pieces of polymer sample in a boiling test tube attached with a thermometer. Keep the tube in the beaker completely stuffed with glass wool. Then heat the beaker gently to melt the polymer sample. After the complete melting of the polymer sample stop heating and allow it to cool. Record the temperature of the sample during the cooling process at regular interval of 30 seconds till to attain the constant temperature. Plot the graph between temperature along Y axis and time in minutes along X axis. The plot as shown in above Fig 5.2 is obtained. Draw the tangent to the curve from both the ends, Draw a horizontal line parallel to time axis (X axis) from the the point of intersection of both the tangents in the region of rubbery state of the sample on the temperature axis (Y axis). Read the temperature on Y axis which is corresponding to the value of glass transition temperature of the polymer sample taken.

Repeat the same procedure thrice and report the average value of observed glass transition temperature of the given polymer sample.

RESULT

The glass transition temperature (Tg) of the given polymer sample by cooling curve method is found to be ... °C.

TABLE 5.0. Showing the Glass transition temperature (Tg) of the various polymers

Sr. No.	Name of Polymer	Tg (°C)	Tm (°C)
1.	Poly vinyl Pyrrolidone	168	
2.	Polytetrafluoroethylene	125	325
3.	Polyethylene (HDPE)	125	135
4.	Polyacrylic acid	106	
5.	Polymethylmethacrylate	105	
6.	Polystyrene	100	240
7.	Polyacrylonitrile	97	
8.	Polyvinyl Chloride	87	227
9.	Polyvinyl Alcohol	85	220
10.	Polyethyleneterephthalate	70	265
11.	Polyethylmethacrylate	65	
12.	Nylon 6	52	225
13.	Nylon 66	50	265
14.	Polyvinyl Acetate	32	

❑❑❑

EXPERIMENT NO. 26

AIM

Determination of glass transition temperature of polymer by Dilatometer method.

APPARATUS

Dilatometer, Temperature controlled oil bath etc.

CHEMICALS

Polymer sample, Mercury.

THEORY

A dilatometer is a scientific instrument that measures volume changes caused by a physical or chemical process (Fig. 5.4). It consists of a glass bulb C, attached with a capillary D graduated in terms volume (cubic mm). The volume of the bulb, up to the zero of capillary D is the net volume of the dilatometer (V_0). The bulb of dilatometer is filled with a sample along with a confining liquid that must satisfy certain requirements. It must exist in the liquid state over the entire range of temperature in which measurements are desired. It must have a vapor pressure sufficiently low so that none of the liquid is lost by evaporation during the course of experiment. Its viscosity must not become unduly high at the lower temperatures. It must have little or no swelling action on the specimen. Its expansivity must be known accurately over the entire range of temperature for which it is to be used. As mercury satisfied all these requirements, it is most commonly used confining liquid.

Water can also be used as satisfactory confined liquid near room temperature, but the relatively small range of temperature between its freezing and boiling points greatly limits its usefulness. There are a number of organic liquids that are suitable for use at low temperatures.

In the determination of Tg by the Volumetric expansion method using dilatometer, the polymer sample is taken in to the bulb of dilatometer and then it is filled with a confining liquid mercury up to the zero mark. The change of volume of the polymer (ΔV) along with the mercury is measured by increase in the mercury level in the capillary tube with the rise in temperature. The plot of ratio of change in volume to initial volume (that is ΔV/Vo) or specific volume along Y axis against the temperature along X axis can be use to determine the value of glass transition temperature Tg of the polymer as shown in the following graph (Fig. 5.3).

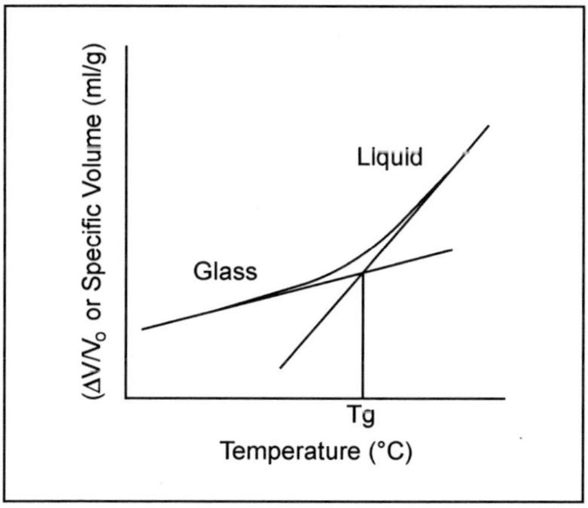

FIG. 5.3. Glass transition temperature by dilatometer method

PROCEDURE

Take the sufficient known quantity of the given sample in the bulb of dilatometer and then fill it with the confined liquid mercury up to the zero mark on capillary. Place the dilatometer in the temperature controlled oil bath so that the bulb of the dilatometer should get completely immersed into the oil bath and heat it. The change in the volume of the polymer is then recorded in terms of the increase in the mercury level in the capillary at different temperatures. The experiment can also be performed by first completely melting the polymer by heating above its melting point and then measuring the change in the volume with temperature during cooling down to room temperature. Note the value of the mercury meniscus level in the capillary for every 3°C increase/decrease in the temperature. Temperature must be controlled to constant value for 5 minutes before the value of the meniscus is noted down.

Plot the graph between temperature along X-axis and the ratio of change in volume to initial volume (that is $\Delta V/V_o$) or specific volume along Y axis. The plot as shown in above fig. 5.3 is obtained. Draw the tangents to the curve from both the ends. Draw a vertical line parallel to Y-axis from the point of intersection of both the tangents. Read the temperature on X-axis which is corresponding to the value of glass transition temperature of the polymer sample taken.

Repeat the same procedure thrice and report the average value of observed glass transition temperature of the given polymer sample.

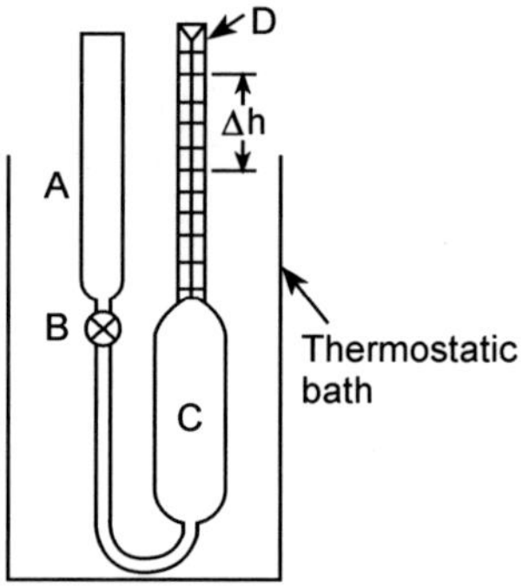

FIG. 5.4. Sketch of the simple dilatometer.

RESULT

The glass transition temperature (Tg) of the given polymer sample by dilatometer method is found to be... °C.

❑❑❑

EXPERIMENT NO. 27

AIM

To study the rate and kinetics of addition polymerization of styrene by dilatometric study.

APPARATUS

Dilatometer, constant temperature bath.

CHEMICALS

Distilled styrene, Benzoyl peroxide.

THEORY

The most convenient method to determine the rate of polymerization is the measurement of the some physical property of the solution that changes with the extent of polymerization. A number of methods are available but dilatometry is the most commonly used method. The method is described below.

The density of a polymer is usually greater than that of the monomer from which it is formed. The rate of addition polymerization can be followed by observing the change in volume of a fixed weight of monomer during its polymerization at constant temperature. The extent of the volume change during polymerization can be recorded by the change in the height of the solution in a narrow capillary tube of the *dilatometer*. The sketch of the simple dilatometer is represented in the figure 5.4. It has the reservoir C to contain the solution which is attached to the capillary D to record the change in volume during the conversion of monomer to polymer with the help of scale marked or the graph paper fixed behind the capillary. The apparatus is to be placed in oil bath thermostat maintained at constant desired temperature of the experiment.

PROCEDURE

Prepare the sufficient solution of freshly distilled sample of styrene monomer containing 0.1% (wt/wt) concentration of initiator benzoyl peroxide. Fill the dilatometer up to lower end of capillary with known volume of this solution. Place the dilatometer in thermostatic oil bath maintaining a temperature of 70°C in such a way that only the capillary tube protrudes from which the change of volume is recorded. In the beginning the volume of solution increases due to thermal expansion till the equilibrium is attained that leads to the increase in the height of solution in the capillary. The overflow of excess solution if occur can be removed by wiping the top of the dilatometer capillary tube using filter paper.

Once the solution attains the thermal equilibrium, the volume of solution will decrease. The height of the solution in capillary should be noted either from the scale on the dilatometer or from the strip of graph paper fixed to the back of the capillary tube at regular interval of time until been completed to prevent the polymer from clogging the capillary. The dilatometer should be washed with acetone carefully and dry using suction.

Repeat the experiment exactly in the same way by taking the 0.2% (wt/wt) concentration of initiator benzoyl peroxide.

OBSERVATIONS

1. Molecular weight of Monomer (M_m) = ...
2. Radius of Dilatometer capillary tube (r) = ... cm
3. Literature value of Density of monomer (ρ_m) = ... g/ml
4. Literature value of Density of polymer (ρ_P) =... g/ml
5. Initial volume of monomer filled in Dilatometer (V_0) = ... cm^3
6. Height of solution in capillary at time t = 0 (h_0) = ... cm (Initial maximum height of solution in capillary tube at equilibrium temperature)
7. Height of solution in capillary at time t = ∞ (h_∞) = ... cm (Last constant height of solution in capillary tube)

CALCULATIONS

1. Initial monomer concentration

$$(M_o) = \frac{\text{Density of monomer } (\rho_m) * 1000}{\text{Molecular weight of monomer } (M_m)}$$

Observation Table

Sr. No.	Time Interval in min.	Height of solution in capillary			$a = h_o - h_\infty = \Delta h_\infty$	Δh_t (x) = $(h_o - h_t)$ cm	(a − x) = $(h_t - h_\infty)$ cm	Log $\left(\frac{a}{a-x}\right)$	Fractional yield of polymerization at time t $(Y) = \frac{(\Delta h_t)}{(\Delta h_\infty)}$	Rate of polymerization at time t $(R_P) = \frac{(\Delta h_t) * M_0}{(\Delta h_\infty) * t}$
		At time t (h_t)	At infinite time (h_∞)	At Zero time (h_o)						
1	2	3	4	5	6	7	8	9	10	11
1.	5									
2.	10									
3.	15									
4.	20									
5.	25									
6.	30									

2. Initial Weight of monomer (W_0 g) = (V_0 cm^3 * ρ_m) g
3. Weight of Polymer at infinite time (W_∞) = Initial Weight of monomer (W_0 g)
4. Volume of Polymer at infinite time (V_∞)

$$\frac{\text{Weight of polymer at infinite time } (W_\infty)}{\text{Density of Polymer } (\rho_p)} = \text{cm}^3$$

5. Fall in liquid height in capillary tube at infinite time ($a = \Delta h_\infty$) = ($h_o - h_\infty$) cm

   Alternatively Δh_∞ can also be calculated by the equation given below if the polymerization does not complete in stipulated time period that is height of liquid column in capillary does not reaches to constant value in stipulated time period.

$$\Delta h_\infty = \frac{(V_0 - V_\infty)}{\pi r^2} \text{ cm}$$

6. Fractional yield of polymerization (Y) at time t is given by the equation,

$$Y = \frac{(h_0 - h_t)}{(h_0 - h_\infty)} = \frac{(\Delta h_t)}{(\Delta h_\infty)}$$

7. Rate of polymerization (R_P) at time t is given by the equation,

$$R_p = \frac{Y * M_0}{t} = \frac{(h_0 - h_t) * M_0}{(h_0 - h_\infty) * t} = \frac{(\Delta h_t) \times M_o}{(\Delta h_\infty) \times t}$$

   Plot the graph of $\text{Log}\left(\frac{a}{a-x}\right)$ against time t, find the slope of the line which is equal to rate constant (K) of the polymerization.

RESULT

1. The rate constant of polymerization reaction is found to be K = ...
2. The Fractional yield of polymerization (Y) = ...
3. The Rate of polymerization (R_P) = ...

❑❑❑

EXPERIMENT NO. 28

AIM

To study the swelling behavior of spherical hydrogel polymers in water.

APPARATUS

Digital electronics balance (with accuracy 0.0001g), Beaker, Glass rod.

CHEMICALS

Given sample of hydrogel polymer beads, De-ionized water.

THEORY

Hydrogels are lightly crosslinked hydrophilic polymers that can absorb and swell in water up to hundreds of times their own weight without dissolving in water. In another word, these are water absorbing natural or synthetic polymers (they may contain over 99% water). When a dry swellable hydrogel comes in contact with water a swelling process starts. The polymer absorbs water and increases its volume. Hydrogels are generally characterized by their ultimate capacity to absorb water (swelling thermodynamics), the rate at which the water is absorbed into their structure (swelling kinetics), as well as their mechanical property in wet or hydrated state (wet strength).

The characteristic properties such as expansion in water (swelling), high water content and elastic nature similar to natural tissues, biocompatibility and lack of toxicity, have enabled them to find extensive applications in biological, medical, novel pharmaceutical area etc. Desirable hydrogel properties for a given application can be achieved by selecting a proper hydrogel material, crosslinking method, as well as processing techniques. These biocompatible materials are currently used in pharmaceutical dosage forms as super disintegrant, ion exchangeable material, and controlled release platform. On the contrary, non-disposable hydrogels with longer term of service have found applications as biomedical inserts and implants. Superporous hydrogels are an exclusive class of hydrogels that can potentially be used for both short- and long term applications including super dis-integrant, controlled release platform, and a gastro- retentive drug delivery system.

An easy way to study the swelling behavior of hydrogels is to disperse the most commonly used and abundantly available spherical hydrogel beads in water and measuring the change in volume, size and weight at regular intervals.

The mass swelling ratio (MSR) denoted by Q_t can be calculated by the following equation

$$Q_t \text{ (MSR)} = \frac{Mt - M_0}{M_0}$$

Where Mt = Mass of hydrogel at any time t.

Mo = Initial mass of hydrogel in its dry state.

When the mass no longer changes with time for a hydrogel in water, it is said to be in chemical equilibrium with the solution in which it is immersed. The parameters called equilibrium swelling ratio Q_e (*ESR*) and the equilibrium water content (*EWC*) are calculated using the following equations,

$$Q_e \text{ (ESR)} = \frac{M_{max} - M_0}{M_0}$$

$$EWC = \frac{M_{max} - M_0}{M_{max}}$$

Where M_{max} = The mass of the hydrogel when it is in perfect equilibrium with the solution in contact. That is the maximum attainable mass of a hydrogel bead.

Water molecules migrate into the polymer network by the process of diffusion. The swelling properties of hydrogels are often found to obey following mathematical relation given below,

$$\frac{Q_t}{Q_e} = Kt^n$$

Where K = swelling constant, characteristic of polymer network.

n = The diffusional exponent characterizing the mechanism of diffusion of the solvent into the polymer network.

t = Time interval of swelling measurement in minutes.

The plot of log $\left(\frac{Q_t}{Q_e}\right)$ along Y-axis as a function of log (t) along X-axis is used to determine the values of n and K from slope and intercept of the straight line respectively.

PROCEDURE

Take a single piece of spherical bead of the hydrogel under study and weigh it accurately by using the digital weighing machine having an accuracy of 0. 1mg. Immerse it in 100ml of distilled water and at the same time start a stop watch to record the swelling time (the time for the sample immersed in water) of the sample. After a definite swelling time t_1, (15 minutes) remove the sample from the water and immediately stop the stop watch. Quickly remove the surface water from the swelled hydrogel with a filter or blotting paper. Take the weight of sample on an electronic balance (0.1 mg accuracy) and the weight denoted as m_1 for the first immersion step. The sample is then again immersed in the water for the second time and start the stop watch. Repeat the procedure until the equilibrium of swelling is reached where the hydrogel bead attain the maximum weight with regular time interval of 15 minutes. At each time t, the weight of the sample is denoted as m_t. The swelling ratio at different swelling time can be calculated by using equations given above. *This method is known as continuous variation method.*

Alternatively number of hydrogel beads of same material having same weight immersed in 100ml of distilled water in separate beakers and labeled them as 1,2,3, and so on. Take the weight of beads at each interval of 15 minutes by removing the hydrogel from the beaker labeled as 1, after 30 minutes by removing the hydrogel from the beaker labeled as 2, after 45 minutes by removing the hydrogel from the beaker labeled as 3 and so on. Remove the surface water from the swelled hydrogel with a filter or blotting paper at each time interval before weighing. *This method is known as separated method.*

OBSERVATIONS

1. M_0 = Initial mass of hydrogel in its dry state. =... g
2. M_{max} = The mass of the hydrogel when it is in perfect equilibrium with the solution in contact. That is the maximum attainable mass of a hydrogel bead =... g
3. Equilibrium swelling ratio (*ESR*)

$$Q_e = \frac{M_{max} - M_o}{M_o} = \ldots$$

OBSERVATION TABLE & CALCULATIONS

For continuous variation method

Sr. No.	Time in Min. (t)	Mass of hydrogel (Mt)	MSR $Q_t = \frac{M_t - M_o}{M_o}$	ESR $Q_e = \frac{M_{max} - M_o}{M_o}$	$\frac{Q_t}{Q_e}$	Log $(\frac{Q_t}{Q_e})$	Log (t)
1.	15						
2.	30						
3.	45						
4.	60						
5.	75						
6.	90						
7.	105						

Plot the graph of log $\left(\frac{Q_t}{Q_e}\right)$ along Y-axis as a function of log (*t*) along X-axis and determine the values of *n* and *K* from slope and intercept of the straight line respectively.

RESULT

The swelling parameters of the given polymer hydrogels are found to be,

1. The Mass Swelling Ratio (MSR) Q_t = ...

2. Equilibrium Swelling Ratio (ESR) Q_e = …
3. Equilibrium Water Content (EWC) = …
4. Swelling constant (from graph, line intercept) K = …
5. The diffusional exponent (from graph, line slope) n = …

❑❑❑

EXPERIMENT NO. 29

AIM

To study the swelling behavior of polymer in different solvents.

APPARATUS

Digital electronics balance (with accuracy 0.0001g), Beaker, Glass rod.

CHEMICALS

Polyacrylamide based Crosslinked polymer sample, Different solvents like water, acetic acid, formic acid, Hexane.

THEORY

Refer experiment number 28.

The swelling coefficient (Q) can be calculated by the following equation,

$$Q = \frac{M - M_O}{M_O} \times \frac{1}{\rho}$$

Where M = Mass of swollen polymer.

M_o = Initial mass of polymer in its dry state.

ρ = Density of the solvent

PROCEDURE

Weigh accurately 1g of given Polyacrylamide based Crosslinked polymer sample. Transfer into the beaker containing sufficient quantity of solvent water. Cover the beaker with watch glass and keep it undisturbed for 48 hours at room temperature. After 48 hours take out the sample from the water and remove the surface water from the swelled polymer sample with a filter or blotting paper. Take the weight of sample on an electronic balance (0.1 mg accuracy).

Repeat the same procedure for different solvents acetic acid, formic acid, hexane etc. Calculate the swelling coefficient of polymer from its initial dry weight (M_0) and swollen weight (M) by using the above equation.

OBSERVATION

Report the observations in following table.

Name of solvent	Density of solvent (ρ)	Initial mass of polymer in its dry state (M_o)	Mass of swollen polymer (*M*)	Swelling coefficient $Q = \frac{M - M_0}{M_0}$
Water Acetic acid Formic acid Hexane				

RESULT

The swelling coefficient (Q) of given polymer sample in different solvents is found to be …

❑❑❑

EXPERIMENT NO. 30

AIM

Determine the amount of formaldehyde in the given solution of formalin and determine percentage of the formaldehyde in the stock solution.

APPARATUS

250ml Conical flask, Pipette, Burette, Volumetric flask, Beaker, etc.

CHEMICALS

Formalin solution, 10% Hydroxylamine hydrochloride solution, 0.2M Sodium hydroxide solution, 0.2 N oxalic acid solution, Methyl orange Indicator.

PROCEDURE

Standardized the sodium hydroxide solution prepared by titrating it against the standard oxalic acid solution using phenolphthalein indicator.

Dilute the given amount of formalin (V_F) solution up to the mark in 250ml volumetric flask by using distilled water (Stock solution). Pipette out 25ml of this stock solution in a conical flask and add 25ml of 10% Hydroxylamine hydrochloride solution to it. Keep the flask as such for 20 minutes with intermediate shaking. Titrate the complete mixture against 0.2M Sodium hydroxide solution using methyl orange indicator. The end point is indicated by the change of colour from red to yellow.

REACTIONS

$$HCH=O + NH_2OH.HCl \rightarrow HCH{=}NOH + HCl + H_2O$$

$$HCl + NaOH \rightarrow NaCl + H_2O$$

OBSERVATIONS

1. Molarity of standardized sodium hydroxide solution = M_{NaOH}

2. Volume of sodium hydroxide required for the Experimental Titration = V_{NaOH} ml

CALCULATIONS

1 Mole of NaOH = 1 Mole of HCHO

1000 ml of 1 M NaOH = 30g of HCHO

1 ml of 1 M NaOH = 0.03 g of HCHO

Hence V_{NaOH} ml of M_{NaOH} NaOH = (?) W g of HCHO in 25ml of solution

W g of HCHO in 25ml of stock solution = V_{NaOH} X M_{NaOH} X 0.03 g

Hence Formaldehyde content in 250ml Stock solution i.e. in given V_F ml of Formalin (Z),

$$Z = (W \times 10) \text{ g}$$

Hence Percentage content of Formaldehyde in formalin solution = $\frac{Z * 100}{V_F}$

RESULT

1. The amount of formaldehyde in the given solution of formalin = ...
2. The percentage of the formaldehyde in formalin solution = ...

❑❑❑

SECTION— IV

References / Bibliography

References/Bibliography

1. *Polymer science* by V.R. Gowarikar, N.V. Viswanathan, Jayadev Sreedhar, New age international (P) limited publication.
2. *Text book of polymer science* by F.W. Billmeyer (Jr.), Third edition, A Wiley-Interscience publication.
3. *Physical chemistry* of polymers by A. Tager, MIR publishers, MOSCOW.
4. *Polymer chemistry* by Seymour and C.E. Carraher (Jr.), Sixth edition, Marcel Dekker Inc., Newyork.
5. *Polymer chemistry and physics of modern material* by J. M.G. Cowie, Second edition, Chapman and hall publication, 1991.
6. *Polymer synthesis and characterization:* A laboratory manual by Stanley R. Sandler, Wolf Karo, Jo-Anne Bonesteel and Eli M.P. Pearce, Elsevier,1998.
7. *A text book on experiments and calculations in engineering chemistry* by S.S. Dara, S. Chand and company limited, New Delhi, 2000.
8. *Introduction to spectroscopy,* by Donald L. Pavia, Gary M. Lampman and George S. Kriz, Third edition, Thomson learning academic resource centre, USA.
9. *Polymeric materials* by S. Radhakrishna and A.K. Arof, Narosa publishing house, 1998.
10. *A laboratory manual of polymers* by S.M. Ashraf, Sharif Ahmad and Ufana Riaz, Second edition, I.K. international publishing house private limited, New Delhi, 2011.
11. *Practicals in polymer science, Synthesis and qualitative and quantitative analysis of macromolecules* by Siddaramaiah, CBS publishers and distributors private limited.

12. Polymer Molecular Weight Analysis by ^{1}H NMR Spectroscopy, *Journal of Chemical Education*, 2011, 88, 1098–1104, ACS publication.
13. Volume Dilatometry by Norman Bekkedahl, *Journal of Research of the National Bureau of Standards*, 1949 (Aug.), 42, 145-156, U.S. Department of Commerce National Bureau of Standards.
14. An experimental study of the swelling behavior of spherical hydrogel in water and the formulation of a theoretical model for its explanation by Pallabi Das, Deeptak Biswas, Dr. Sudipto Roy and Priyadarshi Majumdar, *International journal of drug delivery*, Vol. 4 (1), 2012.
15. A review of polymer dissolution by Beth A. Miller-Chou, Jack L. Koenig, *Progress in polymer science*, 2003, 28, 1223–1270, Elsevier (Science Direct).
16. Modern scientific teaching by R.C. Sharma, 3rd Edition Dhanpat Rai Publishing Company, Delhi.